Praise for *Cowpuppy*

"Hilarious, humble, and heartwarming—get ready for the best true story you will read this year. Follow the adventures of a neuroscientist as he leaves the city behind and finds himself fascinated with the minds, emotions, and souls of a herd that quickly becomes his family. *Cowpuppy* will change the way you think about cows forever. Get ready to fall in love."

—**Dr. Brian Hare**, author of *New York Times* bestseller *The Genius of Dogs*

"In *Cowpuppy*, Gregory Berns shares with us his story of becoming the proud owner of a herd of cattle and the journey into their minds. He provides us a fascinating look into how and why cows and other animals act and react the way they do. I have owned cows for over forty years, but I did not realize how their brains function. Thanks, Gregory, for a very enjoyable read for all."

—**Gabe Brown**, regenerative rancher and author of *Dirt to Soil*

"I am a farmer, but *Cowpuppy* made me feel like a neuroscientist too. It is heartfelt, absorbing, and vivid."

—**Rosamund Young**, bestselling author of *The Secret Life of Cows*

"Combining great and surprising science with deep empathy for animals, *Cowpuppy* is a must read."

—**Tim Flannery**, professor, department of science, University of Melbourne

"A neuroscientist buys a farm, raises cows, and discovers that individual cows have different personalities and a complex social life. You will love his description of how cows react to their reflections in a mirror."

—**Temple Grandin**, author of *Visual Thinking* and *Animals in Translation*

cowpuppy

ALSO BY GREGORY BERNS

How Dogs Love Us: A Neuroscientist and His
Adopted Dog Decode the Canine Brain

What It's Like to Be a Dog: And Other Adventures in Animal Neuroscience

Satisfaction: Sensation Seeking, Novelty, and the
Science of Finding True Fulfillment

Iconoclast: A Neuroscientist Reveals How to Think Differently

The Self Delusion: The New Neuroscience of How We
Invent—and Reinvent—Our Identities

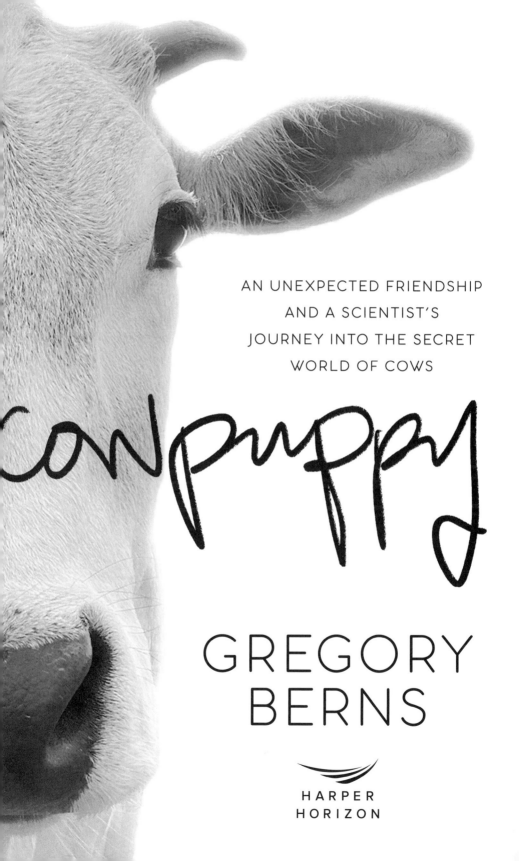

AN UNEXPECTED FRIENDSHIP
AND A SCIENTIST'S
JOURNEY INTO THE SECRET
WORLD OF COWS

cowpuppy

GREGORY
BERNS

HARPER
HORIZON

Published by Harper Horizon, an imprint of HarperCollins Focus LLC.

Any internet addresses, phone numbers, or company or product information printed in this book are offered as a resource and are not intended in any way to be or to imply an endorsement by Harper Horizon, nor does Harper Horizon vouch for the existence, content, or services of these sites, phone numbers, companies, or products beyond the life of this book.

Scripture quotations are taken from the Holy Bible, New International Version®, NIV®. Copyright © 1973, 1978, 1984, 2011 by Biblica, Inc.® Used by permission of Zondervan. All rights reserved worldwide. www.zondervan.com. The "NIV" and "New International Version" are trademarks registered in the United States Patent and Trademark Office by Biblica, Inc.®

ISBN 978-1-4003-3880-1 (HC)
ISBN 978-1-4003-3881-8 (ePub)

Library of Congress Control Number: 2024935632

Printed in the United States of America

24 25 26 27 28 LBC 5 4 3 2 1

For Ken Peek
None of this would have been possible
without your friendship.

Contents

A Note on Cattle Terminology

C attle is the general term for all animals of the genus *Bos*. Cattle are divided into the gendered descriptors *cow*, *heifer*, *bull*, and *steer*. In casual speech, *cow* and *cattle* are used interchangeably as gender neutral, as I do in this book. However, it is never wrong to use the proper terminology:

Cow. An adult female that has had at least one calf.

Heifer. A young female, or an adult female that has not had a calf.

Bull. An adult male with intact testicles. A juvenile is a bull calf.

Steer. A castrated male.

Calf. A juvenile up to one year old. If weaned, it is a weaned calf.

Yearling. A juvenile between one and two years of age.

CHAPTER 1

Princess Xena

Cease, cows, because life is short.

–Gabriel García Márquez, *One Hundred
Years of Solitude* (1970)

The cows were supposed to make life easier. They had but one job on the farm and that was to eat grass. I should have realized that managing a herd of cattle was more complicated than simply putting them out to pasture. Ranchers say you can—and should—eat your mistakes, but that was not my way of thinking. For better or worse, the cows were part of the family.

The calf was the first clue that farming was going to be a lot harder than I ever imagined. She lay beneath the lower fence rail. Her front legs were tucked under her brisket, giving her the appearance of a sphinx. Her emotions, if any, were as inscrutable as her Egyptian counterpart. The momma, a black-and-white miniature zebu, glowered at me while she stood sentry over her calf. The little one had been licked clean and was completely absent of any blood or mucus. As the sun set behind a stand of loblolly pines, it cast a golden aura upon the calf, making her fur shimmer in a beautiful shade of chestnut. She had the most adorable white splotch between her eyes, just like her daddy, a bull we had named Ricky Bobby.

5

I blinked my eyes hard and said to my wife, Kathleen, "Do you see this?" She nodded and said, "What do we do now?"

To say that we were novice farmers would have given us more credit than we were due. On a COVID pandemic–driven impulse, just six months before, we had picked up and moved from Atlanta to a property forty miles south with visions of starting a self-sustaining farm. The cows were the most recent addition to our growing menagerie. I suppose we should have taken the seller's warning more seriously when he delivered the herd and pointed to the black-and-white cow's udder, telling us we would have a surprise soon—a surprise that now lay on the edge of the pasture in the rapidly fading light. But what's a farm without cows? They were going to be the critical element in my grand vision to grow food without the addition of chemicals like fertilizer and herbicides. The cows would graze happily on acres of grass and make manure, which we would compost and turn into the food garden. It would be a glorious cycle of life—or, more precisely, nitrogen. But the cows were becoming much more than mere cogs in the nitrogen cycle. Their cognitive and emotional intelligence was manifesting in ways I hadn't thought possible, forming social bonds that were as strong as the dogs for whom our farm was named.

Talking Dogs Farm was so named because I had spent the last decade of my career as a neuroscientist studying how dogs think. The Dog Project had begun as the half-baked idea of trying to teach my dog Callie to lie still in an MRI scanner so that I could see what she was really thinking. It grew into something much larger, showing how dogs love us, how they see and hear the world, and even which dogs are destined to become good service dogs. I would soon ask these same sorts of questions of the cows.

The calf, though, commanded our immediate attention.

"We can't leave her here," I replied to Kathleen. "The coyotes will get her."

"Call Ken. He'll know what to do."

Pastor Ken and his wife, Boni, were our closest neighbors. Distances in the country were considered on a different scale than in the city. By an urban metric, we didn't have *any* neighbors, at least none that you could see. Out here, *close* meant anybody within a couple of miles. Ken had lived in the area most of his life and had spent his teen years working on a dairy farm. He knew cows. And even though Kathleen and I had moved to the country just six months before, I felt as though I had been here for years and Ken had been my neighbor for as long as I could remember.

Ken picked up right away. I told him about the calf, and he said he would be right over.

As I looked at our new addition, I felt a connection and responsibility rivaling what I had felt when my children were born. The calf made real how much I had changed. I was becoming—had become—a different person. It wasn't just the hard work of farming that was physically transforming me. I was beginning to see my new home in rural Georgia through the eyes of the people on whom I now depended. History ran deep here. So deep that even a conversation about cows and pastures was like an onion, peeling back the layers of everything that had been done to a piece of land and who had done it.

I cradled the calf and carried her toward the barn. The umbilical cord, raw and pink, left a gooey trail on my shirt. I vaguely remembered something about the importance of the mother's first milk, called *colostrum*, and that the calf needed to get that in her gut right away. Instinctively, I felt that calf and momma should remain together undisturbed so that the newborn could get all the milk she wanted without the other cows getting in the way.

The calf must have weighed twenty or thirty pounds, and my clothes were drenched in sweat by the time I got to the barn, just as Ken was pulling up. "Oh, she's a beaut," he crooned. "What a gorgeous little calf."

She replied by letting out a feeble moo that sounded like a wet fart.

With the calf safe in the barn stall, I thought Lucy, her momma, would come right away. Not so. Bellowing in great distress, Lucy kept circling around the area where I'd found her baby. This only agitated the other cows, who began stampeding in circles too.

My daughter, Madeline, home from college, heard the commotion and joined us. With Ken's help, the four of us fanned out into the pasture and slowly began working the cows up toward the barn. It was not like I had seen in the movies. No cowboys or horses or cattle dogs—just four people moving methodically through waist-high grass, applying psychological pressure from a distance to the cows' personal space. It was exhausting. Just when we would get them close to the barn, one of them would spook; the herd would run back to where the calf had been born, and we would have to start all over again.

On about the third try, we coaxed Lucy through the barn gate, which Ken was able to swing shut before the other two cows went through. Ricky

Bobby lowed in protest, but I paid him no attention. The mother cow quickly found her newborn calf in the stall. I closed the door and exhaled in relief. The momma licked her calf furiously.

"You're a cattleman now!" Ken exclaimed.

Only then did I notice that Ken was wearing a back brace. I made a mental note to ask him about that later.

In the morning, momma and calf were exactly where I had left them. I couldn't tell if the calf had moved at all. The temperature was seventy-five degrees and rising on what would be another humid summer day in Georgia. The calf stared at me with dull, sunken eyes.

Kathleen, always the nurse, reminded me that newborn babies experience large fluid shifts and can go from looking like the Michelin tire man to prune creatures in the first day. Still, we hadn't seen the calf nurse, and she was now twelve hours old. The window for colostrum was closing.

The local farm supply sold powdered colostrum. It wasn't as good as the real thing, but not knowing what else to do, Kathleen ran out to pick some up, along with the necessary equipment for administering it. I stayed home and watched YouTube videos on how to bottle-feed a newborn calf.

Cows are ruminants. This means that their digestive systems are highly adapted to extracting nutrients from grasses that other animals can't eat. A horse, although a natural grazer, is not a ruminant and requires food much higher in nutrient quality, like oats and grain and alfalfa. Whereas a horse has one big stomach, a cow's is divided into four compartments, each specialized for a different part of the digestive process. The biggest chamber is called the *rumen* and acts like a giant fermentation tank. After a cow consumes a large quantity of grass, the partially chewed plant material sits in the rumen, where bacteria begin to break it down. Later, when the cow is relaxing, she will burp up the larger pieces, rechew them, mix the material with a large amount of saliva, and swallow the cud again for further breakdown in the rumen. When the food particles are small enough, the second compartment—the *reticulum*—passes the material to the omasum, where water is absorbed. The fourth chamber— the *abomasum*—functions similarly to the human stomach and breaks the food down into components that the small intestine can absorb.

A newborn calf, however, is unable to digest grass or hay. It does not have a functioning rumen because it doesn't have any bacteria in it yet. Like all other mammals, the calf needs its mother's milk to survive and grow, but the complexity of the cow's digestive anatomy means that you can't just give a calf a bottle. If milk goes into the rumen, it won't get absorbed, so it needs to land in the omasum. The path of the food is determined by how the calf holds its head. If the head is tilted up, the milk has a straight shot from mouth to omasum. When the head is tilted down for grazing, food goes into the rumen.

To bottle-feed a calf, you need to tilt the calf's head up and hold the bottle above its head. If that doesn't work, the alternative is to pass a tube down its throat and deliver the milk directly to the omasum. The YouTube videos did not instill any confidence in my ability to do this.

A normal-sized calf weighs sixty to eighty pounds, but because our cows were miniature sized, so was the calf. Kathleen had wisely picked up a bottle and nipple sized for a goat instead of a cow.

The calf stared blankly at me with logy eyes and didn't put up any resistance when I picked her up. Lucy paced nervously.

Kathleen tilted up the calf's head with one hand and jammed the bottle into her mouth. The calf did nothing. Kathleen tried squeezing the nipple, but the milk just dribbled out. The calf just wouldn't suckle.

After an hour of futile attempts to get some milk into the calf, we decided to leave her alone for a bit. Maybe natural instinct would kick in and momma and calf would do their thing. It was also possible that this was Lucy's first calf and she, too, was a bit confused about what was supposed to happen. Either way, time was running out to get some colostrum into the calf. After twenty-four hours, her digestive system wouldn't absorb the crucial antibodies, leaving her wide open to a range of pathogens that could kill her in hours.

We were out of ideas. With nothing to lose, Kathleen put out an SOS on the local Facebook page: *Need help getting newborn calf to nurse.*

There is nothing like an animal in distress to marshal the forces. Within the hour, the page was filled with suggestions and offers of frozen milk. But what we really needed was hands-on instruction by someone who knew what they were doing. There were several cattle farms in the area, so the expertise was here. But the farmers were probably busy tending to their own animals, not monitoring their Facebook feeds.

By early afternoon, the temperature had climbed into the eighties, and the calf was showing obvious signs of dehydration. I called Doc Chanda, one of the two local cow vets, but she was busy and couldn't make a house call. I would have to bring the calf to her if I wanted her help. I didn't even have a truck, and the thought of loading a calf in the back seat of my car seemed like a disaster in the making. We would have to make that decision soon if we were to get the calf to the vet before she closed for the day.

Kathleen's phone rang from an unknown number. Before moving to the country, we would never answer calls unless they were from people in our contact lists. Good thing she picked up this time.

It was the local farrier, a horseshoer named Crowley. Someone up the road, who happened to be having their horses shod, had told him about Kathleen's post. Crowley said he might be able to help.

An hour later, he pulled up in a beat-up truck. He climbed out, stroked his salt-and-pepper beard, and said, "Well, let's see what you got."

The calf was in the same corner she had been since we had put her in the stall. Momma stood balefully, as confused as I was.

"And she hasn't nursed?" Crowley asked.

"Not that we've seen," I said. "And she won't take the bottle."

"Is the cow kicking her off?"

"No," I replied. "She just stands there, unsure what to do."

Crowley nodded. "Lemme give it a try."

I handed him the bottle, which was now sticky with dried milk. "You've done this before?"

Crowley shrugged. "Not with cows, but I have a herd of sheep."

The calf looked at us weakly as we entered the stall. Lucy snorted.

Like a professional wrestler, Crowley put the calf in a headlock and sat down in the straw, backing the calf's rear into the corner so she had no place to go. He lifted his forearm until the calf's muzzle was tipped up at a forty-five-degree angle and jammed the bottle into her mouth.

Both Kathleen and I gasped in amazement as the calf started sucking. It was almost involuntary as the combination of proper head angle and nipple in the mouth triggered the suckling reflex. I could see immediately why our efforts had failed. The calf needed to be in a position that duplicated that of sucking from beneath her momma's udder.

She drained the bottle in less than a minute and actually put up a bit of resistance when Crowley removed it from her mouth. Already her eyes sparked with more life. Crowley spritzed a little of the leftover milk on Lucy's teats and nudged the calf in the direction of her udder. She sniffed a teat and got the milk scent. The calf immediately latched on and began sucking vigorously. She gulped audibly, indicating the milk was flowing.

Lucy's demeanor changed too. The tension seemed to melt out of her body, and she arched her head around and began licking the calf as she nursed.

I was well familiar with the effects of the maternal hormone oxytocin from both medical school and years of studying social bonding in humans and dogs, but never had I seen such a dramatic and instantaneous demonstration of its effects. Oxytocin would have been released in high concentrations during birth, but if the calf hadn't nursed, then the hormone's effects would have waned to the point that Lucy might have rejected her newborn entirely. Then we would have had to bottle-feed the calf for at least six months. In that respect, baby cows are just like baby humans. They need milk every few hours. Thank God we didn't have to go down that road.

Ken popped his head into the stall and beamed at the sight of a newborn calf suckling her momma. "Congratulations to the new father!"

"More like godfather," I corrected.

"True, true." He laughed. "That calf is quite the fighter."

"A warrior," I said as the name dawned on me. "Xena, the warrior princess."

Crowley, his work done, dusted off his pants and headed back to his truck, refusing any payment for his time. He opened the cab door with a creak and paused. "What do you plan to do with the herd?"

I shrugged. "They're for pasture management so I don't have to cut grass."

Crowley shook his head. "They're not dairy cows, and they're not beef cattle. Pretty soon you're going to have a real herd on your hands." He climbed in the truck and added, "Good luck!"

I wasn't sure what he meant by a "real herd," but as I watched Crowley drive away, a heavy feeling settled in my stomach. What *was* I going to do with the cows?

The birth of that first calf changed everything. It set me on a course that would ultimately cause me to see cattle from a whole new perspective. It was also the beginning of a journey that forced me to change my relationship to the land and community that gave sustenance to me and the animals. The cows' role will not likely make sense to most farmers and cattlemen, but maybe by the end of the telling of this journey, all readers—city slickers and country dwellers alike—will see cows in a different light, if only a little bit.

The story begins with how an egghead neuroscientist, who knew nothing about agriculture, ended up on a farm with a bunch of cows.

CHAPTER 2

Buying Tara

Never pass up new experiences,
Scarlett. They enrich the mind.

–Rhett Butler, in Margaret Mitchell,
Gone with the Wind (1936)

In 1969, with the goal of redirecting north- and south-bound trucks around Atlanta, the Georgia Department of Transportation completed construction of a sixty-four-mile-long interstate forming a ring around the city. On maps the new freeway was I-285, but locals called it the Perimeter, which was a curiously militant term that conjured images of Confederates trying to stop Sherman from setting the city on fire in the last months of the Civil War. When Rhett Butler and Scarlett O'Hara crossed the line where the Perimeter would be built a century later, he said, "Take a good look, my dear. You can tell your grandchildren how you watched the Old South disappear."[1]

Now residents of Atlanta are pegged as either ITP or OTP: inside or outside the perimeter. ITP is considered true Atlanta, home to Coca-Cola and CNN; where Martin Luther King Jr. preached; and where the Olympics were held in 1996. OTP is perhaps best known for the old seat of House representative Newt Gingrich. OTP is the domain of soccer moms, boomers, transplants, and refugees. Almost everyone agrees that OTP is decidedly uncool, even the people who live there.

The two major north-south interstates, I-75 and I-85, form an X in the center of the Perimeter. They merge for a stretch of ten miles, forming a perpetually congealed mass of traffic called the Connector, before splitting again north and south of downtown. This arrangement of interstates neatly divides Atlanta and its environs into four sectors corresponding to the points of the compass. In recent decades, most of the demographic expansion had been to the north and east. The northern suburbs were largely white and upscale. They had spread almost fifty miles north to the foothills of the Appalachians. The climate there turned out to be rather conducive to viticulture, and several wineries were doing a steady business. Apple and blueberry picking thrived too.

In contrast, the eastern sector was where the international transplants and refugees settled. The International Rescue Committee was headquartered in this area and had become the landing hub for people from all over the world seeking a better life. Communities from every country in Africa dotted the east wall of the Perimeter. This was ironic seeing as how the most famous landmark there was the gigantic Confederate mural of Jefferson Davis, Robert E. Lee, and Stonewall Jackson that was carved into Stone Mountain. Every few years some Atlantan politician would float the idea of sandblasting it into oblivion, but the idea would inevitably go nowhere.

The western sector was a no-man's-land. Once you passed Six Flags, you were pretty much in Alabama. And to a Georgian, no state was the butt of more jokes than Alabama. The only redeeming feature of Alabama was Talladega, home of the biggest, baddest NASCAR event in the country.

That left the south, whose most noteworthy feature was Hartsfield-Jackson International Airport, the world's busiest airport. If you lived in Atlanta, this was a real gem because you could fly direct to almost anywhere in the world, but for everyone else, it was a stopover in hell. The linear layout of the terminals meant a sometimes impossibly long traverse between connecting flights. South beyond the airport, though, the next major attraction wasn't even in Georgia. It was the Gulf of Mexico—aka the Redneck Riviera.

Since the late 1990s, Kathleen and I had lived fifteen miles straight north of downtown, which placed us just OTP. This was stereotypical suburbia. Unpretentious homes were nestled on half-acre lots cut neatly out of the

pine forests. The southern pines grew up to two feet a year in the subtropical climate, and since the first wave of suburban neighborhoods was built in the 1970s, the trees towered at one hundred feet. That seemed to be their limit. The pines grew vertically so fast that they didn't have time to develop much width and ended up looking like very large bottle brushes. When Kathleen and I moved to Atlanta, I was struck by how big the trees were. Besides the great redwood forest in California, I hadn't seen anything like them. The pines, though, didn't have the longevity of the redwoods. Eventually a pine would become top-heavy and tip over in a strong wind. If it was in a stand of trees, the fallen timber would just lean heavily on its neighbors. But if it was a lone pine, it would fall to the ground, often taking a powerline with it. I had gotten used to these monthly power outages and finally procured a generator for just such events.

Apart from living with a spotty power grid, life was comfortable. The demographic expansion of Atlanta had resulted in a restaurant scene that rivaled those of New York City and Los Angeles in its culinary breadth. The dog-scanning project I was working on continued to roll out new findings about the canine mind every year.

COVID-19 changed all that. The university shut down and went virtual. I had assumed this would be a temporary pause and didn't worry too much about carrying on the dog research. But as the first year of the pandemic stretched into the second, it became harder and harder to train dogs for the MRI. We tried training over Zoom, but the dogs needed hands-on practice. Masks made everything more difficult. Dogs rely on visual cues from their owners as much as verbal ones, and the masks obscured both. The last straw came when I had to turn down a contract from the Department of Homeland Security to study how dogs process different types of odors. In the chaos of the pandemic, I knew we wouldn't be able to make the milestones, and the program officer refused to extend them.

With all the time I was spending in front of the computer, I started trawling for properties. It filled idle time and fed into a growing *Green Acres* fantasy of giving up the city life for the country. I could Zoom in from a farm just as well as from our suburban home. The only catch was that I knew nothing about agriculture, but that was easily remedied. The explosion of in-home learning meant that I could be a student just as easily as a teacher. I signed up for Farming 101 and Vegetable Production through the Cornell

Small Farms Program. Oregon State offered a rigorous twelve-week course in permaculture.

A little knowledge is a dangerous thing. From what I learned from the experts teaching these courses, the farming life seemed doable. I had made it through medical school and residency, and farming couldn't be harder than that, or so I thought. The main problem was an economic one. While growing food seemed within my reach, making a living doing so seemed impossible. Even the instructors, who were all expert farmers, admitted they needed a source of off-farm income—like teaching courses in how to farm. Indeed, the first lesson was that farming is a business. If you expect to succeed, you better have a business plan. Food production was one thing, but selling it was something entirely different.

Kathleen and I went through the exercises leading up to a mission statement for our fantasy farm. This is the kernel of any business plan. Our mission statement stated who we were and what we wanted to accomplish. The first exercise asked, "What three words describe your farm?" Seeing as we didn't yet have a farm, we had to imagine what kind of farm we would want. There would be animals, of course. And I imagined food production in a way that improved the environment. We settled on these three concepts: *self-sustaining, happy animals, healthy.*

The next question was, "What promise of quality do you commit to providing your customers?" It seemed unlikely that as novice farmers we would be able to grow food that was better than any other local farm, but we could say this: *By consuming our products, you are making the planet a better place for all living things.*

But here was a conundrum, faced by every farmer throughout all of time. Food is a living thing. So how can we make the planet better for all living things if we need to kill some to eat? I knew we would not be in the business of raising animals for slaughter, so our imaginary farm would have to be limited to fruits and vegetables. But what's a farm without animals? These contortions of logic led to a rather aspirational mission statement: *We produce food and fiber using self-sustaining agriculture over a multigenerational timescale, while also drawing down carbon from the atmosphere. We treat the animals on our farm as partners in the endeavor, not to be exploited.*

Highfalutin to be sure. But if you don't aim high, then you will never know how far you can reach.

It was a crisp winter day in Georgia, late 2020. Kathleen was in the passenger seat staring impassively ahead. I couldn't read her expression, but her silence betrayed her emotions. She always clammed up when she was unhappy, usually about something I had done. And this most certainly was my doing.

We crossed the south wall of the Perimeter, and I wondered if we had made a huge mistake. All I could see were strip malls stretching to the horizon. The stores were of the lowest caliber: fast-food franchises, but not the good ones, and what seemed to be an excessive number of storefronts for auto supplies. In twenty-two years of living in Atlanta, I had never had cause to visit this part of the metropolis. Now it was clear why. And we were considering moving here? I said nothing, for nothing would have helped make my case.

Heading south, ten miles outside the Perimeter, I lost count of the number of businesses we had passed whose main offering was some sort of fried chicken product. Although there were sidewalks in front of some of the stores, the distances between establishments offered no safe passage for pedestrians. But who would be crazy enough to walk along a four-lane road with a speed limit of fifty-five anyway? The answer was obvious: only those who had no choice.

Eventually the continuous string of strip malls began to break up. Expanses of forest, a mile at a stretch, filled the spaces between the low-slung prefabs. And then we passed through some sort of invisible barrier, for then there were no more stores.

Forests gave way not to mini-malls but to farmland. My mood brightened at the sight of horses roaming a pasture. And not just a gentleman's pasture like you might see in the tonier suburbs of Atlanta. No, this was the real deal. Acres and acres, so big that I had no sense of scale. I couldn't say whether these pastures were ten or a hundred acres. Giant hay bales, rolled into tight cylinders, dotted the fields. A hand-painted sign, nailed to a mailbox at the end of a dirt drive winding into the woods, read Fresh Eggs $4.

Kathleen saw it, too, and I could feel the tension in the car drop a notch. Still, I said nothing. This was all new to us.

According to the 2010 US Census, 47 percent of the census blocks have a population of zero.[2] When you live in a city, it is easy to be unaware that nearly half of the country is essentially uninhabited, but that was exactly what was before us. Land. Forest. Pasture. Only the occasional house gave away that there were, in fact, people here, but the density was minuscule. No subdivisions. No tract homes. The closest thing to a mixed-use development was a country store attached to a home.

From the road, our destination was unremarkable. A three-rail white fence ran alongside the road, enclosing a pasture with a pair of horses grazing contentedly. The driveway passed through a wrought iron gate, and when we drove through, a pastoral vista opened before us. Four additional separate pastures housed a bevy of horses. These were sleek athletes, not the old nags I grew up trail riding. One of them raced the car along the fence line as we drove up toward the main house.

We passed beneath a tunnel of crepe myrtles and found ourselves on a gravel circle in front of a house that could only be described as Tara. It was a white two-story affair, large but not obscenely so. Massive white columns formed a portico on the front facade, and the second-story veranda caused me to look over my shoulder for anyone who hadn't heard the Civil War was over. This was antebellum architecture in all its glory. Not really my style, but the rest of the property took my breath away. The words that sprang to mind were *God's country*.

The real estate agent, a cheery blonde woman with a smooth Georgia accent, greeted us with a smile and ushered us through a modest front door into an expansive central hall. If it had been closed off at one end, it would have been called a foyer, but it extended all the way to the rear of the house, giving the hall the feel of a central lobby. A set of pocket doors opened into a large dining room with a baby grand piano. A pastoral mural had been painted on all four walls. In one panel a Confederate soldier, one arm in a sling, the other dragging a musket, was returning home to his wife and two children working in the fields. Farther down the hall, a cozy wood-paneled office housed floor-to-ceiling bookshelves. A portrait of Robert E. Lee hung on the wall. Kathleen and I exchanged wary glances, both of us thinking we had wandered into some kind of *Dukes of Hazzard* universe. Setting aside the owners' historical predilections, the house was arranged in such a way that it drew you in. The main hall ended in the great room, which combined an open kitchen and den next to a wood-burning stove. A screened porch on the rear of the house invited you to enjoy the sounds of the forest without the torment of bugs.

The house, beautiful and comforting as it was, was not the main reason we had journeyed to middle Georgia. We had come for the land and all the potential it held. We piled into a golf cart for the grand tour. Next up was the stable, which was nearly a hundred feet long. Inside, six stalls lined a central corridor. Rubber mats covered the concrete floor, sparing the horses undue stress on their hooves. I had grown up with horses, but this was beyond anything I was familiar with. Jokingly, I asked if the horses came with the stable. The agent laughed. She didn't have to say it, but the horses were worth more than the entire property.

Around the rear of the stable, gravestones rose up at random angles from the forest floor. There seemed to be two distinct clusters of graves. The larger group held twenty or thirty plots, most dating between the late 1800s and mid-1900s. The other group of graves was cordoned off with a low brick wall and wrought iron fence. I recognized the name from the property records. This was the family plot of the people who built the original house. The man had died twenty years before, but the woman passed away only three years ago. They were buried next to their son, who, according to his headstone, died at the age of fifteen. Just outside the family plot, there were three hand-made markers for their dogs. It gave me the heebie-jeebies.

We continued past the cemetery, paralleling the fence line of the largest of the pastures. Two horses lazed in the center of the field, trying to catch some warmth from the bright December sun. They paid us no attention as we drove past them and into a forest. A path meandered through a stand of pines a hundred feet tall and then down an incline toward a creek. We crossed over a ford, where a pipe emerged horizontally from the berm, trickling water into the creek. A spring.

Up the hill on the other side of the creek was the workshop. A tractor and a backhoe were parked outside.

"Are the tractors included?" I asked.

The agent smiled. "Everything is negotiable." She knew I was hooked.

We passed through another stand of pines, bigger than the first, and now the path was covered in pine straw. Back in the city, we paid good money for the stuff, using it as mulch around the yard. The golf cart bounced over some ruts as we exited the pine forest into a meadow of clover. The agent didn't waste any time here, though. She zipped through, dodging ant mounds the size of footballs. It was a good thing she didn't stop, because these were fire ant mounds. On the other side of the meadow, we plunged back into forest. Instead of pines, hardwoods towered over us. Oak, maple, and the ubiquitous sweetgum, despised for its seedpods that looked and felt like spikey golf balls.

The agent pulled up when the ground became too soggy to continue. She pointed through the woods. "The river is down there."

It was a good thing I had brought my hiking boots. I hopped out and began trudging through the mud to see if I could find the edge of the property. The Flint River, sister to the more famous Chattahoochee, was one of the two major rivers in Atlanta. While the Chattahoochee originated in the Appalachians in northeast Georgia, the Flint's headwaters were under Hartsfield. From there, it meandered lazily south, picking up volume, until it merged with the Chattahoochee in far southwest Georgia. From there, the confluence emptied into the Gulf of Mexico. The Flint was one of only forty rivers in the entire United States that flowed unimpeded for over two hundred miles.

Fittingly, Tara abutted the Flint in *Gone with the Wind*. Margaret Mitchell wrote, "The muddy Flint River, running silently between walls of pine and water oak covered with tangled vines, wrapped about Gerald's new land like a curving arm and embraced it on two sides."[3]

Even though Tara was fictional, Mitchell described exactly what I saw. Water oaks—with their small leaves—towered over me, and indeed, vines snaked up their trunks. A slough prevented direct access to the river, but I sought higher ground until I could work my way around the muck and reach the bank. The Flint oozed lazily beneath an old train trestle, which looked to have been built from pines harvested on-site. Way up on the underside, someone in decades past had painted Pink Floyd Lives.

There was nothing, really, to think about. Tara had seduced me. I could already see my future there. Lush green pastures with some sort of grazing animal, maybe goats, and rows of vegetables that we would sell at the local farmer's market. On the way home, I looked at Kathleen and asked, "What do you think?"

We had been together for over thirty years, and in that span of time, you come to know a person's ways of thinking. Kathleen was not a person to make impulsive decisions. She liked to turn things over in her mind until she settled on a course of action. But the magical thing about relationships is the limitless capacity of one person to surprise another, if only occasionally.

Kat did not even hesitate. "Let's do it."

CHAPTER 3
The Cows Come Home

I could dance with you till the cows
come home. Better still, I'll dance with
the cows, and you come home.

—Groucho Marx, *Duck Soup* (1933)

I was not normally one given to mystical experiences, but I couldn't shake the feeling that a force—something just out of sensory perception—had brought us to Tara. Maybe it was the tranquility, or maybe it was the abundant natural resources, but there was something undeniably magical about the place. Perhaps that was why the Muscogee Creek tribe had made it their home centuries ago.

When the keys were handed over to us, the seller gave us his notes about the property. Much of the handbook was devoted to practical knowledge of things like wells and septic tanks, but he had also noted the remnants of a fishing weir on the river, where the Creek had built a crude dam to funnel fish. The provenance of the land was ever present. Each morning I would step outside and watch the sun rise over the eastern pines and cast its rays on the mist hugging the pastures. Signs of civilization were distant, but I could feel the presence of all those who had once called this home. Now it was my turn to be a good steward of the land.

Without any horses, the grass in the pastures had grown at an alarming rate by May. No doubt this was related to the fertilizer that I had put down the month before. In hindsight this was probably unnecessary, but I had assumed that that was what farmers did. And this might have been true had I been growing actual crops or raising livestock, but since neither had yet occurred, the grass seemed to be knee-high all the time. I swear if I sat there long enough I could see it creeping skyward.

So I began mowing.

I'd fire up the John Deere, which, even though it put out a modest forty-four horsepower, seemed massive compared to the little push mower I had used all my life as a suburban home dweller. I would back the big green tractor up to the mowing implement, called a Bush Hog, being ever so careful to line the three-point hitch to the pins on the mower. This usually took several back-and-forths. Once aligned, I jimmied the drive shaft of the mower onto the spline of the power take-off (PTO) device on the rear of the tractor. The PTO scared the bejeezus out of me. Although the tractor was of recent vintage and therefore had safety features that prevented the PTO from turning when nobody was sitting in the driver's seat, it still looked like the arm-and-hair mangler that it had been in years gone by. I usually turned the engine off when I hooked up the PTO.

With Bush Hog in tow, I would set to mowing the pasture, trying to make neat, straight traverses of the paddocks, like all the neighbors did. The trick Zamboni drivers use on hockey rinks also works for mowing pastures. First, you cut the perimeter. Then, on the next pass, you make a cut lengwise down the center. You loop back to the side you started on and come back down the far side of the center cut. If you do it right, you end up with perfectly straight rows without doubling back or having to go in reverse.

There is a Zen-like aspect to mowing fields, especially if you don't mind the smell of diesel. The thrum of the engine masks all external sounds, while the constant vibration lulls one into a state of semiconsciousness. Some call it "diesel therapy," which I take as code for escaping the wife and kids.

The previous owner of the farm had built a lean-to for storing the tractor and keeping it out of the rain, but he had neglected to make it tall enough to park the tractor with the canopy attached. The Stetson I had bought to keep the sun off my head did just that but nothing else. After an hour or two of mowing, with the heat of the diesel blowing in my face and the Georgia sun beating down, I had usually had enough. With a six-foot Bush Hog, it took an hour to mow one acre. I had twelve acres of pasture to cover. Every week.

By late May, Georgia had begun to settle into the heat the South was known for. With plenty of time to ponder my situation, diesel therapy lost its appeal. It didn't take long to realize that I needed some biological help.

It may seem obvious that a pasture is for the keeping of animals—more specifically, grazing animals. As the previous owners of Tara were horse breeders, the pastures had been designed for that purpose, ranging in size from one to five acres. A riding arena had even been carved out of the biggest one, although the jumps had been removed. A neat oval of fine-grained sand remained, as if a UFO had picked up all the apparatuses en masse.

Although a pasture is for grazing, it is the fence that defines it as a pasture. Without a fence, a pasture becomes a field, or, if you're out West, a range. The act of enclosing a swath landmarks it—and everything contained within—as owned. The choice of fencing, then, says something about what is being kept inside, and, by the same token, what is being kept out. Fence maintenance is probably the least desirable chore on a farm, but it is a necessary one, for even a single gap has the potential to negate a fence's entire purpose. Fence condition also speaks volumes about the person who erected it and maintains it (or not). A straight fence with tight rails? You can bet the farmer is on top of his game. Posts rotted and tipping over, with slack barbwire? Overworked and probably looking for an exit ramp.

All five of our pastures were easily identified as those of the equine variety. Each was enclosed by a three-rail wooden fence. Four-by-six posts had been sunk exactly eight feet on center, with each two-by-six rail carefully leveled. Everything was painted dark brown, giving the property an elegance that broadcast how well the horses had been taken care of. Someone had put a lot of work into building these, and now it was my responsibility to maintain them. Already I could see dozens of rotted rails that needed replacing. The whole thing could have used a new coat of paint, as well.

When I was a young kid growing up in Southern California, my family had kept two horses in the backyard, and we had similar fencing. The horses were purely for trail riding, but mostly I remember spending a lot of time shoveling out the stalls and feeding the horses copious amounts of oats and hay. And then there were all the accoutrements: saddles, bridles, halters, currycombs, and a cornucopia of balms and ointments to keep the flies at bay. Even with all the attention showered on them, they seemed to be surprisingly fragile. Feed them the wrong thing or look at them the wrong way, and they would develop colic. I have vivid memories of my parents walking the gelding in circles all night to prevent him from lying down and twisting up his colon.

Sure, horses were fun to ride, but the upkeep and expense seemed unlikely to counterbalance any benefits from their grazing the pastures.

Goats seemed more in line with what I thought a farm animal should be. They are intelligent and known to have distinct, almost doglike, personalities. Goats struck me as perhaps the most natural evolution of my research on the canine mind. But goats are also reputed to be troublesome escape artists, and the existing pasture fencing would do nothing to contain them. A goat would just walk between the rails or jump over them. Sheep, although less impish than goats, would also flow through the fencing, as if it were nonexistent. Plus, sheep were easy pickings for the coyotes that prowled the river.

I had moved to the farm partly so I could have any animal I wanted, but the existing fencing was going to dictate what animals I could keep. It came down to horse-sized animals: creatures too big to squeeze between the fence rails but not so big they could break them down.

The only animals left on my list were cows.

Although ranching wasn't a major industry in the area around Tara, within a few miles there were a few farms that kept cattle. Without knowing exactly what I was looking for, I started cruising the back roads, ogling the bovines. One farm had three Guernseys that always seemed to be lazing under a big oak tree by the road. But that was the exception. Every other cattle farm had a herd of Black Angus. What I had in mind was what most people thought of: a gentle black-and-white Holstein dairy cow. As luck would have it, a dairy farm was located just a few miles away.

A rusted sign, Rucks Dairy Farm, hung forlornly over the complex. As I pulled up, the sole farmhand gave me a welcoming wave, even though I was an uninvited stranger. I introduced myself as a new neighbor and told him that I was looking to get a few head, preferably a steer or two.

Given the signage, I was expecting dairy cows, but all I saw were Black Angus. To my untrained eye, they were big and imposing in a rather homogeneous manner. Because they were all black, it was hard to discern any differences from a distance, even though I knew there must be a range of personalities and physical attributes. On the plus side, all Angus cattle are polled, meaning they didn't have horns.

The farmhand shook his head and explained that when the tornado ripped through the farm in 2011, Rucks decided to get out of the dairy business. Now they just raised Angus cows and sold them at auction. If I was willing to pay the going rate for pounds on the hoof, then a deal could probably be made.

I ran the numbers in my head. The live-auction rate was running about $1.50 per pound, and I didn't see any cattle that looked to be less than one thousand pounds. That put the price at $1,500 a head. I said I'd have to think about it.

Although many people keep a single family cow, usually for milk, my instinct told me that if I was going to keep cows, I needed at least three to constitute a small herd. In fact, I would need more than that if they were to act as biological mowing machines for twelve acres. But if I had to pay the live-auction rate for full-sized cows, this would become a very expensive proposition. All that to avoid mowing? No, I needed something smaller, like a weaned calf.

Where does one go to buy some cows? Most ranchers would go to the local sale barn. And, indeed, there was an auction every Wednesday. But without an expert to guide me, I wouldn't be able to tell a good, hardy cow from a downer. Besides, the auctions were cesspools of disease passing between the animals. Never mind that I didn't have a truck or trailer to bring home any animals. The only alternatives were classified ads.

Although goats and chickens dominated the listings, there were plenty of people selling cattle on the farm section of Craigslist. Some were liquidating entire herds, probably because it was too hard to make money in cattle ranching. There were ads for weaned calves and a few for cow/calf pairs. And lots of bulls. It seemed like everyone was trying to unload a bull. *Yes*, I thought, *it's tough to be a male on a farm, especially one with testicles.* A listing for miniature zebus caught my eye. I didn't know what exactly a zebu was, but miniature sounded good.

Turns out, zebu is a generic term for the Asian breeds of cattle. Commonly known as Brahman, they have a large hump between their shoulders and are well adapted to hot climates. A miniature zebu is just a smaller version. In India they are called *nadudana*, which means "small cattle" in Hindi. At one time, miniature zebus were bred to be kept on temple grounds.[1]

The ad was for a bull and two cows. The seller wanted seven hundred dollars for each. I didn't really want a bull, but I made arrangements to check them out, hoping that we could negotiate for individuals.

Kathleen gave me a skeptical look when I told her that I had found some cows to check out. "What are you looking for?" she asked.

I didn't know the first thing about cows, let alone how to tell if it was a good deal. I did, however, assume that like everything on Craigslist, the seller was trying to get rid of a problem. I replied, "I guess just see if they're healthy."

Kathleen rolled her eyes. As if I really knew what a healthy cow looked like.

The cows were located ninety minutes away, in a northeast suburb of Atlanta. Ironically, they were not too far from where we had just moved from. We traveled through subdivision after subdivision until suddenly we were at a "farm." It was obvious that the owner had lived there for a long time, probably refusing to sell out to developers. The result was a rural oasis in a desert of tract homes. I wondered how he got around the inevitable zoning laws about livestock.

The farmer was exactly what I expected given the surroundings. He was an older gentleman, puttering around with a tall walking stick that could have been used as an ambulatory aid or an implement to round up the animals, or both. He greeted Kathleen and me warmly and introduced us to his menagerie.

Chickens ran freely, scratching the dirt here and there. Two goats jockeyed for position on an elevated platform, while another pair played with a pile of sticks. A horse stood in the far corner of the corral, gazing at the lush green grass on the other side of the fence. And three cows huddled together, wary of the strangers that had just driven up.

The bull was the most striking of the three. He had a white underbelly,

but his head and back were varying shades of chestnut brown and black, giving the impression that someone had poured earth-toned paints down his spine. A hump jutted up between his shoulders, like the horn on a western saddle. A pair of testicles dangled impressively between his hind legs. It almost goes without saying that it is very important for a bull to have two testicles. Undescended testicles are generally infertile and prone to cancer. No problem here. This bull had a healthy swagger that exuded gentle confidence. The other animals gave him the deference he commanded, without him being a jerk about it.

The two cows had decidedly less personality. The females nosed around an empty hayrack, even though it was early spring and the grass was coming up nicely. They tagged along behind the bull, following his lead without showing any inclination for exploring on their own. One was a mix of gray and white and the other was mostly black with a band of tarnished silver circling her midsection. She looked almost like the famed belted Galloways of Scotland.

All three appeared healthy and well-nourished. I couldn't see the ribs on any of them. The females' hip bones poked out above their hindquarters, but their spine bones weren't visible. Everyone's noses and eyes looked clean. No discharge or flies buzzing around them.

"They look good," I said to the old man. "How long have you had them?"

"'Bout a year, I reckon."

I asked him why he was selling them.

The man shrugged and turned toward his house. "Wife only allows me a certain number of animals." He aimed the top of his walking stick at the cows, adding, "Plus, both cows are pregnant."

This was unexpected. If he was fixing to make a profit, surely he would wait until the calves were born. Then he could sell five cows instead of three. I wondered if there was something wrong with the animals. Buying livestock was turning out to be like buying a used car. Caveat emptor. But maybe the old man's explanation was true. If his wife was after him about having too many animals, then he really would want to unload them before he had two more.

"When are they due?" I asked.

"Dunno. Have to check my calendar." He squinted at the two cows. "Soon, I s'pose."

I mulled over the situation. Each of the three appeared to weigh no more than four hundred pounds. Imposing animals to be sure, but nowhere near the Angus cattle I had seen. If I was going to wade into the world of cattle, this seemed like a good bunch to start with. I looked at Kathleen.

She shrugged. "It's up to you. This is your passion."

I asked if we could get a closer look.

The man picked up an empty three-pound coffee container. "I use this to lure them into the stall." The bull caught the movement from the corner of his eye and altered his ambling course from the pile of sticks to a stall where the man was headed. The cows followed along without much excitement. On the way, the man filled the can with some pellets. The sound of rattling grain got the cows' attention, and their laconic pace immediately transformed into a lively trot.

When the cows were penned up in the stall, we got a good look. They appeared as healthy up close as they did from a distance. The bull ducked his head down and waggled it a bit, shoving his way between the two cows. He came up to the man and extended his tongue. It was kind of endearing. The man obliged the bull's plaintive request by pouring the grain into a trough.

"Are they friendly?" I asked.

"The bull is."

I didn't know if that meant the females were unfriendly or just indifferent. I suspected the latter. In any case, it was time to make a deal. These three seemed like as good a group as any to begin our adventure.

"If you can bring them to us," I said, "you have a deal."

A week later, the old man delivered on his promise, pulling his truck up to the barn with a small red animal trailer in tow. He still had his walking stick with him, now looking more like a shepherd. With his free hand, he opened the tailgate and let the herd free to discover their new home.

The cows didn't hesitate. The bull poked his nose out and, sensing freedom, jumped down, followed closely by the white cow and then the black one.

"Hey, look," the man said, pointing his staff at one of the females. "You see that blacker cow right there?"

"Uh-huh," Kathleen said.

"Well, I was gonna tell you 'bout the surprise."

"Surprise?"

"Yeah. Her udder just done dropped."

It didn't take an expert to know what that meant. She would be calving soon. How soon? Hard to say. Some cows "bagged up" a couple of hours before calving. Others, a couple of weeks.

Dairy cows often get names, but as a general rule, ranchers do not name their cattle. This is understandable, as beef cattle are eventually headed to freezer camp. And although a prolific bull might be lucky enough to get a moniker, we humans have to deal with the cognitive dissonance resulting from the obvious fact that that luscious slab of prime rib came from a living creature. Being nameless goes a long way toward distancing us from the animal.

Our little herd, though, was neither beef nor dairy. I knew they would never be sent to freezer camp. Not on my watch. Of course they needed names. The pair of females, who seemed to be friends, at least most of the time, brought to mind Lucy and Ethel from the old TV show *I Love Lucy*. It wasn't clear which one should be Lucy and which Ethel, so, rather arbitrarily, the black one became Lucy and the white one Ethel. With this naming scheme, the bull had to be Ricky (for Ricky Ricardo).

It is amazing how much personality you can discern just from observation of animals' behavior patterns.

Like the old man said, Ricky was the friendliest. I had picked up a fifty-pound bag of cattle cubes at Tractor Supply Company—my new favorite store. The "cubes" were actually one-inch-long cylinders of compressed alfalfa, and I had begun feeding them by the handful as treats to get the cows comfortable with me. Ricky wasted no time in gobbling them from my hand, stretching out his tongue like some sort of prehensile hand. Kathleen thought it was gross, but the dexterity with which he used his tongue reminded me of an elephant's trunk. Ricky would tilt up his head and extend his tongue in probing movements, trying to take all the cubes out of my hand. But I meted them out, only dropping them into his mouth if he let me scratch him behind the ears or stroke the big wattle under his chin, called the *dewlap*.

Maybe it was the tongue action, but I could see he had a sort of goofy personality. Easygoing at times, but then it was pedal-to-the-metal if food was on the line, which reminded me of a quote by Will Ferrell's character, Ricky Bobby, in *Talladega Nights*: "If you ain't first, you're last."[2]

And so Ricky Ricardo became Ricky Bobby.

While Ricky Bobby seemed to warm to his new environment, Lucy and Ethel were another story. I couldn't get within six feet of Lucy. If I stretched out a hand, she would just turn away. I decided to leave her alone. Maybe that was just how she was, or maybe she needed time to get her bearings. While Lucy was not ready to accept me, Ethel seemed a bundle of conflicts. When I fed Ricky Bobby, she would hold back a length or two, staring at me wide-eyed. I got the distinct impression that she desperately wanted treats, too, but was afraid to approach. The hair whorl on her forehead was positioned at high noon, almost dead center between her horns, and when she stared at me with her big brown eyes, the whorl beamed at me like the Eye of Providence on the back of a one-dollar bill. Temple Grandin, a professor at Colorado State University and the world's most famous authority on cattle, has written that the higher the whorl, the more high-strung the cow.[3] If so, I had a live wire on my hands. It made me wonder what was going on in Ethel's brain.

CHAPTER 4

Cow Brains

McCoy: We just want to talk to somebody about Spock's brain. That's all.
Kara: Brain and brain! What is brain? It is Controller, is it not?

—"Spock's Brain," *Star Trek* (1968)

In one of the most ridiculous episodes of *Star Trek*, Spock's brain is stolen by a race of aliens that have all their needs met by a centralized computer called the Controller. When the Controller begins to show signs of age, it magically beams the female leader, Kara, aboard the *Enterprise* to abscond with Spock's brain as a replacement. McCoy and Kirk go to retrieve it and put it back in Spock's body. The episode is widely regarded as the worst of the series, but precisely because of its groanworthiness, it is a must-see for fans. Still, it makes you think about what a brain can (and can't) do.

Why do animals have brains?[1] The simplest answer is because they need to survive and reproduce. Without being too glib, these are really the only two things an animal has to do in the world. But what an animal needs to do, and how it does it, depends on both its physical form and the environment in which it is situated. As a result of these physical and environmental differences, animals' brains are as different from one another as the animals themselves. These variations aren't random. The Darwinian law of nature suggests that these differences reflect the varied adaptations a species has gone through to survive. It is why the brain of a carnivore looks different from one of a ruminant, like a cow.

Until recently, neuroscientists had focused on the brains of a handful of species: Humans, of course, because we want to understand how we think; mice and rats because they are plentiful and because invasive procedures can be done to them that can't be done to humans; and monkeys because they are primates, like us, but not quite as humanlike as chimpanzees, which now are protected from most invasive research. And that's pretty much it. Considering there are over six thousand different species of mammals, we haven't even scratched the surface of brain diversity in the world.[2]

As a neuroscientist, I have always believed that the study of the brain can reveal things about the mind that are unknowable by psychological approaches. Magnetic resonance imaging (MRI), in particular, has let us peer inside the skull and see the brain in action. Decades of these sorts of experiments have begun to disentangle the complex relationship between neural activity and the subjective experiences we have. In 2011 I began training dogs to lie still in the MRI scanner so we could use the same tools to figure out how their brains functioned. Over the next decade, we learned that there are many similarities between dog and human brains. For example, we both have reward systems responsive to primary rewards, such as food, and more complex rewards, like social praise.[3] And, of course, there are differences, notably in how we process human speech. Humans are symbolic. We label everything and understand that these labels are symbolic representations of things in the real world. Dogs do not have the neural mass for such a complex function.[4]

Apart from dogs, it seems unlikely that any other animal can be coaxed into an MRI scanner voluntarily. As tame as my cows were, crawling into a narrow tube and being subjected to the jackhammering of the MRI was not in their repertoire of behaviors. And I was not about to sedate them just to get a look at their brains.

In the quest to understand the minds and brains of the other species on this planet, a few neuroscientists—myself included—had begun accumulating a database of brains by scanning the brains of animals that had died.[5] Good structural information can be acquired if the brain is extracted and preserved within forty-eight hours of death. It sounds morbid, but if an animal has died or been euthanized, what better way to bring meaning to the life they lived?

Cows, though, had not been a high priority in neuroscience research. The lack of information only furthered the common misperception that cows are dumb animals. It was time to set the record straight on cow brains.

As a general rule, big animals have big brains. While this may seem obvious, the reasons for this relationship continue to be debated. It's not because bigger animals have more muscles. True, their muscles are larger, but the number of muscles that need to be controlled is surprisingly constant across the animal kingdom. What seems to dictate the size of the brain is the surface area of the animal. The more surface area, the more sensory input that needs to be processed.[6] Because surface area is proportional to volume raised to the two-thirds power, the weight of an animal's brain is also proportional to the weight of the body raised to the two-thirds power.

This brain-body weight relationship holds, on average, across all mammals, but the ratio displayed in an individual species may be more or less than this rule suggests. The encephalization quotient (EQ) measures the degree to which a species' brain is bigger or smaller than would be expected for its size. Cats, for example, have an EQ of 1.0, meaning their brain size is exactly average for their body size. Dogs fare better with an EQ of 1.2. Monkeys and chimpanzees have EQs of about 2; the bottlenose dolphin's EQ is 4; and humans have an EQ of about 7. Cows, though, are below average, with an EQ of 0.6.[7]

Ever since the brain-body relationship was described in the 1970s, EQ has been taken as a neural proxy for intelligence. Brains consume a lot of energy, so what is all the excess neural tissue for if not smarts? The problem is there is no standardized test of intelligence that works across species. In recent years, the A-not-B test (ABT) has come the closest to serving as an intelligence metric. As originally conceived by the developmental psychologist Jean Piaget, the ABT placed two boxes in front of an infant. The experimenter would place a toy under box A and then, after a brief delay, unveil the toy, delighting the infant. The activity would continue in this fashion for several trials, during which infants would often begin reaching toward the box to get the toy. The experimenter would then make a crucial switch and hide the toy under box B. Infants younger than about ten months would continue to reach toward the A location, despite clearly watching the tester place the toy under the other box. By twelve months, almost all children would correctly switch to box B. Correct performance indicated an understanding of where the toy was located, what Piaget called *object permanence*.[8]

The ABT is easily adapted to animals, replacing the toy with a morsel of food. In 2014 researchers reported on the relationship between brain size and performance on the ABT in thirty-six species.[9] They covered a wide range of animals, from birds to monkeys and chimpanzees to dogs and elephants (although no ruminants). Surprisingly, EQ had only a weak relationship to ABT performance. Notwithstanding the elephants, who couldn't do the ABT, overall brain size was the best predictor of performance.

As you might expect, there is a wide range of brain sizes across the animal world. An adult human brain weighs fourteen hundred grams—about three pounds. The brain of another intelligent animal—the bottlenose dolphin—weighs one thousand grams. Man's best friend—the dog—has a brain that weighs about one hundred grams, although this ranges from fifty to one hundred twenty grams, depending on the breed. The cow clocks in with a respectable four hundred grams (almost one pound). Size, though, is just the crudest overall measure of brain complexity.

In the decades I had spent studying both human and animal brains, I'd scanned the brains of more than one thousand people and, during the Dog Project, over one hundred dogs. I'd scanned the brains of dolphins, sea lions, coyotes, and even an extinct marsupial, the Tasmanian tiger. If there was one thing I'd learned, it was this: Brains are as individual as the creatures to which they belong. Even within the same species, every brain is different.

Given such a wide range of sizes and shapes, it is reasonable to assume that these variations in brain morphology reflect something about function. Animals that depend on smell as a primary sense have relatively more of their brains devoted to olfaction. Visual creatures, like humans, have large portions of their brains devoted to the processing of visual information. The relative sizes of different parts of the brain and how they're connected to each other can reveal a lot about the perceptual world of other animals.

Apart from size, the overall shape of a brain stands out as an obvious difference. The dog's brain is shaped like a fat torpedo, whereas the dolphin's is spherical, like a soccer ball. In profile, the human brain looks a bit like a clam. The cow is intermediate between the canine and human forms. The overall shape of a brain is primarily a reflection of the animal's skull. Canids have pointy heads and so have pointy brains, except for dog breeds that have been bred to have shortened muzzles. Pugs, Boston terriers, and French bulldogs all have roundish brains. The shape of the cow's brain reflects its skull shape, which, although elongated, is also broad and pie-plate flat between the eyes.

The next obvious difference is in how brains are folded. Unlike size and overall shape, the degree of folding, which is called *gyrification*, indicates how much neural tissue has been crammed into the skull. Think of the brain as a crumpled-up sheet of paper. This is not just an analogy. The nervous system is one of the first organs to form in an embryo, and it begins as a flat sheet of cells that curls up into a tube. As the embryo develops, the tube grows a pouch at one end that will become the brain, while the rest becomes the spinal cord. The brain continues growing, but because it is encased by what will become the skull, the brain has to fold in on itself.

The more folded a brain, the more neurons it contains.[10] The neurons are contained in a sheet about three millimeters thick. This is the gray matter. The connections between neurons constitute the white matter. To a first approximation, mammalian brains fold up in a stereotypical pattern. This initial folding gives rise to the major lobes of the brain, with the biggest

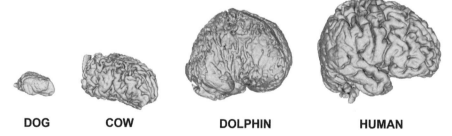

DOG   **COW**   **DOLPHIN**   **HUMAN**

Renderings of the brains of different animals. The dog, cow, and dolphin brains were scanned with MRI after being removed from deceased animals. Because they were positioned upside down for scanning, the tops appear flattened. (Cow data courtesy of Yaniv Assaf.)

fold separating the temporal lobe from the frontal and parietal lobes. After that, continued folding allows the sheet of neurons to continue expanding. The dog's brain, modest in size, does not contain a large number of folds (although it is more folded than the cat's brain). At the other end of the spectrum, the human brain is highly folded, which is how it is able to hold eighty billion neurons. The dolphin's brain is folded to a high degree, too, but its folds are lacelike, which may help regulate its temperature in cold water.[11]

What about the cow brain? It has a rather impressive amount of folding. In fact, the amount of folding is comparable to primates with similarly sized brains.[12] Does that mean a cow is as smart as a monkey? Maybe, but few researchers have even tried to measure the intelligence of cattle.

To learn more about the cow's brain and what makes the animal tick, we have to metaphorically and literally go beneath the surface. Coming by a cow brain, though, is difficult in the United States. The FDA has long banned the consumption of cattle materials they consider at high risk for bovine spongiform encephalopathy—mad cow disease. This includes all parts of the nervous system. Also, because most cattle are stunned with a bolt to the brain before slaughtering, there isn't anything left for an intrepid researcher to study. Luckily, Yaniv Assaf, a neuroscientist at Tel Aviv University who has been scanning the brains of a wide range of animals, was kind enough to send me his MRI scans of a cow brain. With this data, I was able to render the surface for comparison to other animals and peer inside.

The interiors of mammalian brains tend to look more similar than the exteriors. These internal parts, which contain the brainstem and the limbic system (aka the lizard brain), are thought by many researchers to be the parts of the nervous system that evolved first, long before mammals came on the scene two hundred million years ago. The exterior sheet, or neocortex, came later.

Although I've never liked the term *limbic system*, it is useful shorthand for a set of interior structures that straddle the brainstem. The limbic system is critical for olfaction, memory, and emotion. Although portions of the limbic system are located in the cortex, most of its components are subcortical, meaning beneath the cortex. These include the amygdala, which is critical for high arousal emotions, especially fear; the hippocampus, which coordinates memory formation; and parts of the caudate nucleus that are involved in reward and motivation. When the cortex is stripped away, we can

Cowpuppy

see these components. Whereas the overall size of the cow's brain is moderate, the subcortical parts of the limbic system are about the same size or larger than those in humans' brains. For example, both the cow and human amygdalae are about 1,200 cubic millimeters. The cow hippocampus is 4,300 cubic millimeters, whereas the human's is 3,500 cubic millimeters.

Because a cow's limbic system is as big as or bigger than a human's, it makes up a larger proportion of the total brain weight. This is more a reflection of the expansion of the cortex in humans for all of our cognitive functions, especially language and abstract thought. But the similarity of limbic systems also tells us that cows possess the neural architecture to experience emotions, probably very similarly to humans. And the size of the hippocampus shows how important memory is in the cow's life. There are different types of memory, but spatial memory is a major function of the cow's hippocampus, allowing it to remember where things are located and to find its way in complex foraging environments. The large amygdala, which is a sort of hair trigger for the brain's alarm system, is consistent with the cow's flighty predisposition. At the other end of the emotional spectrum, the nucleus accumbens, which sits at the bottom of the caudate, collects information about what is likely to be rewarding in the cow's environment. By coordinating the release of dopamine, the accumbens acts as an accelerator of motivation to seek out things a cow finds pleasurable. Food, of course, but social rewards too. Just like in humans and dogs.

One might wonder whether the structural similarity between bovine and human brains represents something special about cows, or whether all domesticated livestock share these traits. The answer appears to be a bit of both. Certainly all livestock have well-developed limbic systems, with parts sized comparably to the cow's. Without taking anything away from the cow, we can assume that the other barnyard animals also have highly developed brain structures for emotion and memory.

Where the cows distinguish themselves is in cortical folding. It turns out that all the ungulates—hooved four-legged mammals—have highly folded cortices. This is a characteristic of the order *Artiodactyla*, which includes both even-toed ungulates and their distant water-living *Cetacean* cousins, dolphins and whales. In terms of farm animals, this includes cattle, pigs, sheep, goats, llamas, and alpacas. But even within this group, the cow stands out, scoring higher on gyrification than all but the llama.[13] Horses are in the order *Perissodactyla*, which are the odd-toed ungulates and include the zebra, donkey, and rhinoceros. They, too, have highly gyrified brains, even a bit more than the cow.

Gyrification is correlated with intelligence, presumably because more folding allows for more neurons to be crammed into a small volume.[14] The ungulates have a thinner cortical sheet, which allows for easier folding during development. *Cetaceans* have the thinnest cortical sheet of all the mammals and also have the most gyrified cortex. And while it is tempting to conclude that the high level of gyrification in dolphins is related to their intelligence, we simply don't know if the relationship between gyrification and intelligence holds across orders. Within an order, though, gyrification may well predict an animal's intelligence. If so, horses may be the Einsteins of the farm, but cows come in a close second.

Intelligence is one thing, but the brain doesn't reveal anything about how animals comport themselves. In other words, their personalities. For that, you need to take a close look at how each individual behaves.

CHAPTER 5

Cowpuppy

Happiness is a warm puppy.

–Charles M. Schulz, *Happiness Is a Warm Puppy* (1962)

Even before Xena was born, the cows' personalities had begun to reveal themselves. As the old man said, Ricky Bobby was the friendliest, although by friendly I think he meant that Ricky Bobby wasn't afraid of strangers, as Lucy and Ethel were. The precise term is *flight distance*, which measures how close you can get to an animal before it runs away. As long as I kept my hands where he could see them, Ricky Bobby had zero flight distance. If I approached slowly, I could walk right up to him and scratch him behind his ears. Ethel had a flight distance of about six feet, while Lucy's was double that. When I fed them cattle cubes, Ricky Bobby would take them right out of my hands, but Lucy and Ethel would eat them only if I put the treats on the ground and backed away.

Lucy lightened up a bit after Xena was born. I could tell by her body language. When Lucy first arrived at the farm, she wouldn't even make eye contact with me. If I stared too intently, she would avert her head. And if I tried to approach, she would arch her body away, positioning herself so that she always had an escape route, her muscles rippling with tension. But after Xena came on the scene, Lucy's tension diminished. It wasn't a dramatic change in behavior, but I could see it in the lessening of stress in her body.

This was unexpected, as I had assumed that she would have become even more standoffish to protect her calf. But Pastor Ken wasn't surprised. After we had helped Xena latch on that first day, Ken was convinced Lucy understood that we were not only not a threat but also her friends. Maybe so, or maybe it was just her getting used to us with time. Did cows have friends? Ranchers have long observed that cattle tend to form cliques and hang around with the same individuals. I wondered whether cows viewed humans similarly, as friends.

On about her third day of life, Xena got over the shock of birth, and her personality began to emerge. She did not appear to possess any fear, and she had somehow not inherited her mother's expansive personal space. Xena readily approached me and sniffed around curiously. She was a cheerful, easygoing calf. Irresistibly cute, she would bat her big doe-eyes with lashes that women would die for. When nursing, she would raise her tail like a flag and wag it back and forth as a dog might.

Of course, all newborn animals are irresistibly cute. Biologists theorize that cuteness evolved to elicit nurturing responses in parents, although we humans tend to be the most exuberant in extending this reaction to other animals. As Xena was my first calf, I had no others for comparison. It was possible that all newborn calves were like her, although I doubted it. All animals exhibit a range of personalities, and cows should not be any different. The question is, at what age do they begin to differentiate from each other? More importantly, how does one go about describing animal personality?

Without more data points, it was hard to characterize Xena's personality. Comparisons to the adults would do no good because they had already had years to develop their personalities. And who knew what travails they had been through? I didn't have to wait long, though. Xena's sibling arrived three weeks later.

With one calf under my belt, I had a better idea of how to look for impending signs of calving. Ethel showed no signs of her udder dropping, but bagging up was not a reliable indicator of when calving would occur. She did, however, *spring* and begin trailing mucus down her rump. These were signs that her birth canal was getting ready. Sure enough, two days later, she separated herself from the herd and paced around a remote corner of the big pasture.

I didn't want to hover and stress her further, so I fell back to checking on her from a distance with binoculars. I still missed the actual event. At the 9:00 a.m. check, Ethel was pacing. An hour later, a small white animal was wriggling in the grass beneath her legs. The grass was so tall, I almost missed it.

We went through the same routine with Ethel's calf as we had with Lucy's, confining momma and calf in the barn stall to let them bond without being pestered by the rest of the herd. Unlike Xena, though, the new calf had no trouble nursing. She latched onto the teat right away and began sucking with loud slurps and gulps. I dipped the end of the umbilical cord in iodine, as was standard practice for a newborn. I didn't see any testicles and told Ken we now had a second heifer.

In the morning, I let momma and calf out of the stall. Lucy and Ricky Bobby galloped over to check out the new herd mate. The calf was still very wobbly and was no match for Ricky Bobby, who kept bunting her around the corral. I couldn't tell if he was imprinting on her scent or telling her that he was the boss, or both. It did not appear to be an act of aggression, as his head butts were relatively gentle, more like nudges. He hadn't acted this way when Xena was born. I wasn't sure why he was doing this with the new calf, but he clearly understood the size differential and didn't want to hurt her. Even so, he kept pushing the calf between the fence rails. Naturally, this caused Ethel great distress. She lowed loudly, unable to get to her calf, who couldn't figure out that she could simply crawl back through.

Ricky Bobby soon lost interest in the calf. He turned his attention to the pasture and ambled off to graze on the grass, heavy with morning dew. Lucy followed with Xena in tow, while Ethel hung back a bit in case Ricky Bobby continued his bullying behavior.

Throughout the day, the calf kept ending up on the wrong side of the fence. Unsure if this was the calf's doing or Ricky Bobby's, I checked on the herd every few hours and, on several occasions, found Ethel and the calf running back and forth on opposite sides of the fence. Fortunately, the calf showed no fear of me and let me scoop her up to shove her back through the rails. She struck me as a rather joyful, carefree animal, to the point of being a bit unaware of her surroundings. She reminded me of Lisa Kudrow's character on *Friends*.

So Ethel's calf became Phoebe.

Since the calves weren't mature enough to graze, all of their nutrition came from their mothers' milk. The increased metabolic demand of producing milk meant that the mommas had to eat as much grass as they could. So while the cows were busy grazing much of the day, the two calves had a lot of time to occupy themselves. Xena and Phoebe bonded with each other and became fast friends. Ricky Bobby contented himself with spending the days lazing in the middle of the pasture, chewing his cud while keeping a watchful eye over his brood.

When the calves weren't nursing, they were usually in close proximity to each other. During the day, they nosed around the pasture, taking in the sights and smells. They would copy the adults and nibble on some grass now and again, but their digestive systems weren't ready, and they would end up spitting it out.

Xena and Phoebe soon learned the rhythm of the herd's day. It was August and the days were long, so the herd was in no particular hurry to get up at first light. By about 7:00 a.m., the mommas rose from wherever the herd had decided to bed down. They stretched their legs and exercised their bodily functions with massive cow patties and copious amounts of urine. With the lanes cleared, the calves latched onto their mommas and worked their way around the four teats of the udders. After ten or fifteen minutes, the calves were sated, and the adults began their morning graze. This lasted until noon, when everyone would plop down and relax to chew their cud. With no cud of their own, the calves curled up with each other to take a nap.

Cows have an interesting way of lying down. You might think that cows are like horses, who sleep standing up, but that is not the case at all. Like dogs, cows actually spend a lot of time lying down. When cows lie down, they first kneel on the equivalent of their elbows and then plop down their rears. Their front legs end up tucked beneath the brisket. The sequence is reversed when cows stand: rump up first and then a quick rocking forward as they straighten out the front legs.

The other endearing quality about how cows relax is that they will some-times roll on their sides. The first time I saw them do this, I thought they were ill. But they were just very relaxed. Xena and Phoebe especially liked to sleep this way. I often spied the two of them lying flat out, one resting her head on the other. Because this was apparently a natural behavior for them, I figured that if they were comfortable enough with me, they might someday cuddle with me, using my lap as a head rest.

When Phoebe was about three weeks old, I noticed a change in her underside. Something had swelled between her hind legs. She was sprouting testicles! Phoebe was, in fact, a little bull calf. I felt stupid, but now everything made sense. The testicles were undescended at birth, which was why I had missed them. But how could I have missed the wiener? Well, it was pretty small and tucked up in a fold of skin close to the belly. I just hadn't paid close enough attention to where exactly the urine came from when he peed.

It also explained why Ricky Bobby had been so intent on bunting the calf around the pasture. He knew the calf was a future bull and needed to make clear who was the boss.

Sheepishly, I told Kathleen that I had mis-sexed Phoebe and that she was actually a boy.

"I guess we have to change her name," she said.

The cows had not shown any sign that they knew their names, but on the off chance they did, I figured we should pick something that sounded similar. "How about BB?"

"For B. B. King?" Kathleen asked.

"No," I replied. "Ball bearings."

By early September, even though the days were still uncomfortably hot, the number of daylight hours had begun their inexorable slide toward winter. The herd adapted by beginning their morning graze earlier. The other change, which wasn't yet visible but I knew was coming, had to do with the grass. The herd had been content to graze the large pasture, a lush five-acre expanse of bermudagrass. This would have been enough to feed a herd of a dozen miniature cows, but bermudagrass goes dormant in the winter, turning into what amounts to brown straw. This would not sustain the cows through winter, so I had purchased a hundred bales of hay from a farmer on the other side of the county.

You can feed hay simply by placing the bales on the ground, though that will result in a lot of wastage. Cows are picky eaters, but they are also sloppy ones. They will take a mouthful of hay bigger than they can swallow, dropping clumps out of their mouths as they chew. They will trample the hay on the ground, walking right over a fresh bale, and then they will pee and poop on it. Once any excrement touches the hay, they won't eat it. Without a method to keep the hay contained, up to 50 percent of it will get scattered and wasted. The solution is a hayrack, which keeps the hay contained in one place and prevents the cows from walking on it.

For reasons of frugality, and because part of the farming ethos was to be self-sufficient, I decided to construct my own hayrack rather than purchase one. The main advantage of building my own hayrack was that I could size it for miniature cows. The frame was constructed from scrap two-by-fours, to which I attached two pieces of cattle fencing, forming a *V* in which to place the hay. I put off deploying the hayrack as long as possible, forcing the herd to graze the pasture up until the grass looked inedible, which was about the first week in October.

BB was three months old and Xena four, and their personalities had begun to diverge in interesting ways. Although my initial impression of Xena was that of a friendly, curious calf, as she got older she became wily and more suspicious, a bit like her momma. It wasn't a dramatic change. Rather, it was more the fact that BB went in the other direction, becoming even more curious and happy-go-lucky. If I hadn't had BB for comparison, I would have thought that Xena was just as friendly as the day she was born.

Characterizing animal personalities is a tricky task. Academics are persnickety about it and draw a distinction between temperament and personality. Temperament is considered to be an inherited set of tendencies that appear in infancy and form the foundation of personality, which only emerges through time and experience.[1] While most researchers would agree that animals have temperaments, many think that only humans have personalities. This is partly a practical limitation of how personality is measured. In humans, questionnaires are used to probe how an individual reacts and feels in different circumstances. Because an animal can't answer questions about how it feels, we must infer personality traits from behavior patterns,

Cowpuppy

which is, in essence, temperament. Another reason academics are hesitant to apply the term *personality* to animals is that it anthropomorphizes them. If an animal has a personality, it makes it harder to justify harming them, whether for laboratory research or for consumption. Nobody wants to think about the personality of the cow that was turned into a standing rib roast.

Anybody who has lived with animals, though, knows that they have personalities. Over time, it is easy to observe how a pet reacts to different situations. It doesn't take long to see that every animal has its own idiosyncrasies. Some dogs, for example, are gregarious and unafraid to approach strangers with tail wagging. Others hold back, approaching only when they've decided it is safe, and others still are constitutionally afraid of strangers, running away or barking in fear. While my academic colleagues would characterize these traits as animal temperaments, I think the distinction from personality is unnecessary. What matters is describing animal personality in a consistent way.

Researchers have come up with several different systems for characterizing human personalities, although they all tend to be variants of the descriptions that the ancient Greeks used. Hippocrates thought that four bodily fluids controlled human personality. These four *humors*—blood, yellow bile, black bile, and phlegm—gave rise to the four major personality types. (In other cultures, the four humors correspond to the four seasons or the four elements: air, fire, earth, and water.) We might call these personality types *sanguine* (enthusiastic and social), *choleric* (aggressive and short-tempered), *melancholic* (depressive), and *phlegmatic* (reserved).

Of course, there are more than four types of people. Modern personality theory posits that the four humors are different dimensions of personality that exist to varying degrees in each person. The most common framework adds a dimension and is called the *five-factor model of personality*, or just the Big Five.[2] Each factor ranges on a continuum between two opposites. The N factor stands for neuroticism and ranges from neuroticism to emotional stability. The A factor is for agreeableness versus antagonism. E is for extroversion versus introversion. O is for openness or closedness to new experiences. And C is for conscientiousness versus impulsiveness.

There is growing evidence that the Big Five—or their equivalents—exist in nonhuman animals too.[3] Some factors seem to translate well to animals while others need some modification. Openness in animals can manifest as curiosity and exploratory behavior. Similarly, it is easy to tell whether an animal is an extrovert or an introvert by how they react to new people and animals. Agreeableness, too, seems to translate to animals by the manner in which they comport themselves with each other. Some are friendly and easy to get along with while others are jerks. In humans, neuroticism is characterized by anxiety, depression, and moodiness, but it is difficult to apply those terms to animals. It is easier to describe an animal in terms of activity versus inactivity, the latter being a sign of depression. Conscientiousness translates the least well, so we substitute a dimension of boldness versus shyness. A sixth dimension—dominance versus submissiveness—is often added. However, because dominance also depends on an animal's herd or packmates, it may not be a stable personality factor.

To recap, the Big Five for animals are the following:

1. **Aggressiveness:** aggressive versus peaceful
2. **Openness:** curious versus uncurious
3. **Boldness:** bold versus shy
4. **Activity:** active versus inactive
5. **Extroversion:** extrovert versus introvert

So how do we measure these dimensions? The most objective method is to create some sort of test. In cattle, researchers have measured reactions to provocations like touching the cow from head to toe or the approach of a strange human, and how quickly a cow flees after being released from a restraint chute.[4] While such probes have the advantage of objectivity, they are cumbersome to implement and require putting multiple cows through the test under conditions that are held constant. Even with this type of data in hand, it must still be translated into the Big Five dimensions.

Alternatively, we can assess a cow qualitatively on each of the Big Five dimensions. This approach depends on the person doing the assessment and their experience observing cattle. Although my cattle experience was effectively limited to the five cows inhabiting my pastures, I could still assess them relative to each other. All I had to do was rank the five cows on each dimension, and then I would have a personality fingerprint for each of them.

Why did I want to do this? Partly for my own curiosity. Although I had had the cows for only four months, it was obvious that each had a different personality. Cattle ranchers talk about difficult cows, primarily as a justification to remove them from their herd, but surely, I thought, there is more to cattle than being easy or difficult. Another reason had to do with animal husbandry. My herd was multiplying and would continue to do so. Some of the personality dimensions might be heritable. Maybe I could be more nuanced in the breeding of personality traits than simply selecting for easy or difficult.

When I constructed graphs of the five dimensions, an interesting pattern of personality fingerprints became evident for the herd. Lucy ranked the lowest on most of the dimensions, reflecting the fact that she wasn't terribly extroverted and could be a bit of a bully sometimes. Mostly, though, she kept to the business of being a cow, eating grass, and nursing Xena. She was what most people had in mind when they thought of a cow. Ricky Bobby ranked high on many of the dimensions, resulting in an expansive ring on his personality graph. He was, of course, the bull and acted the part, scoring the highest on aggressiveness. This was situationally dependent, as he wasn't aggressive very often, but still, none of the other cows acted the way he did.

Personality fingerprints of the herd using the Big Five.

BB had the most interesting personality profile. Like Ricky Bobby's, it was an expansive pattern, with the exception of the aggressiveness dimension, leading to a bowl-shaped fingerprint. BB had a doglike personality: extroverted but not aggressive; curious and open to new experiences; bold and active.

He was a cow but so much like a puppy. He was a *cowpuppy*.

BB's curiosity, although endearing, sometimes got him into trouble. You know what they say about curiosity and the cat, and BB seemed determined to prove that the adage applied to cows too. When I put the hayrack out for the first time, BB gobbled up the hay, just like the other cows. But for reasons only the cows knew, he was at the bottom of the social hierarchy, which meant that he had last choice of where to get access to the hay. While the adults lined up perpendicular to the hayrack, BB was left to feed from the narrow ends where the panels formed a *V*.

BB didn't seem to mind, that is, until he put his whole head into the crook of the *V*. When he pulled back with a mouthful of hay, he didn't understand

BB eating from the hayrack. Both calves are beginning to mature, developing humps on their backs. Shortly after this, BB got his head stuck in the crook of the V at the end. (Gregory Berns)

that he needed to lift his head up before pulling back. And so he got stuck. As soon as he felt pressure on his jawbone, he panicked and started pulling even harder, digging his feet into the ground to gain leverage. This, of course, only made things worse. Even though he didn't weigh more than sixty pounds, he was panicking so much that it took all my strength to push his head in enough to free him.

Up until this point, I had mostly left the cows alone in the evening. But after witnessing BB's inadvertent attempt to hang himself, I had a feeling in my gut that I should check on him to make sure he wasn't stuck again. It was a good thing I did. Not an hour later, BB had done exactly the same thing. His head was wedged in the V, and in his attempts to escape he was beginning to drag the hayrack out to pasture. The other cows milled about, unable to help. Ethel lowed plaintively, knowing her calf was in distress.

In the morning I added cattle panels to the ends of the hayrack so BB couldn't get his head inside. But from that point on, the evening check became part of the daily routine, which forced me to take a closer look at what I was feeding the cows. Although I didn't realize it at the time, this was also a pivotal moment, for it was the first step toward my becoming a part of the herd.

CHAPTER 6

Grass

I celebrate myself, and sing myself,
And what I assume you shall assume,
For every atom belonging to me as
good belongs to you.
I loafe and invite my soul,
I lean and loafe at my ease observing
a spear of summer grass.

–Walt Whitman, "Song of Myself," *Leaves of Grass* (1855)

anguage scholars can't decide on the exact origin of the word *grass*. In Old English, they called it *græs*, from the Proto-Germanic *grasan*. In Old Norse, Dutch, and German, it was simply *gras*. Some think that *grass* is related to *grow* because of the Proto-Indo-European root, *ghros,* for young shoot or sprout; or *ghre*, which meant to become green. Grow. Green. Grass. You can see how they're related phonetically and thematically.

In the first year, I couldn't tell the difference between one kind of grass and another. The whole point of the cows, after all, was to eat the damn stuff so I wouldn't have to spend so much time on the tractor mowing it. But of course it wasn't that simple. Before the cows, the grass was just another part of the overall farm maintenance. The cows changed that. After they arrived, the grass wasn't just something to look at and mow when it got too tall. Grass was the cows' lifeblood.

It was ironic, then, that I had procured the cows to manage the pastures, but their very existence forced me to learn how to manage the grass. They hadn't made life easier, and on account of the time I spent with them or thinking about them, the cows had made my life demonstrably more difficult. If the universe contained an uber-consciousness, surely it took pleasure in these twists of fate. I had to laugh at the irony and adapt.

Although fresh grass is desirable to raise cattle, it is not strictly necessary. It is possible to run cattle entirely on a mixture of hay and prepared feed, like grains. Industrialized feedlots are engineered for hundreds of cows to belly up to feed troughs, which is why they're called *concentrated animal feeding operations* (CAFOs). Even ranchers that run grazing operations will use supplements, called *rations*, to make up for nutrient deficiencies in their forage or to boost calories so their cattle put on weight as fast as possible. From a production standpoint, the rancher needs to acquire the cheapest input in the form of the cow's ration to maximize growth rate. Big-ag companies are happy to sell the stuff too. But I didn't get the cows to feed them rations. They had a job to do.

That didn't mean they would do the job the way I wanted them to. Animals are like workers: they need management. And for that I had to take some control over what was growing in the pastures to incentivize the cows to graze where the grass needed cutting and to stay away from the areas where the grass was too short.

I had to become a grazier.

The origin of the word *grazier* is likely related to the origin of the word *grass*. As a note of pronunciation, cattle are *grazers*, with the z sounding like that of *blazer*, but *grazier* is pronounced like the name *Frasier*. The phonetically related *grease* comes from the thirteenth-century Anglo-French *grece*, meaning the oily fat of land animals. Later, in French, it became *graissier*, meaning someone who fattened livestock. And a *greaser* was someone who smeared salve on sheep. In any case, the modern grazier is someone who pastures animals for market, a job that centers on managing the growth cycles of both the grass and the animals that eat it.

Grass belongs to the plant family *Poaceae*, which contains about twelve thousand species. A few dozen of these are familiar to us as the grasses that make up suburban lawns, but the thousands of other species include the cereal grasses like wheat, rice, corn, and barley, as well as the native grasslands that cover the prairies of North America and the savannahs of Africa. Bamboo is also a type of grass. Grasses are the most important source of food on the planet for humans. Direct consumption of grains alone accounts for half of all human dietary energy.[1] Indirectly, grasses account for even more through their consumption by livestock that eventually end up on the dinner table.

A field of grass is made up of thousands of individual plants, each with an anatomy characteristic of the *Poaceae* family. An individual plant is a collection of shoots called *tillers* that originates from its base. A tiller can arise from the base of an existing plant, or it can shoot up from a horizontal extension. When the extension is above ground, it is called a *stolon*; below ground it is a *rhizome*. Leaves are attached to the tiller at *nodes*. As the plant grows, the space between nodes (the *internode*) elongates. Leaves only grow from the nodes. This adaptation resulting from the plants' symbiotic relationship to grazing animals allows the ends of the leaves to be grazed without inhibiting the plants' growth. As a tiller grows upward, it also grows horizontally, sprouting daughter plants through stolons and rhizomes. As long as the tillers remain connected to the mother plant, they can share reserves of energy.

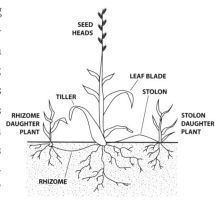

Components of grass.

Grass has three stages of development in its life cycle. In the first, the vegetative stage, the plant sprouts new leaves from the nodes. This begins at the node closest to the ground, which is called the *crown*. A field in the vegetative stage is short like a carpet but beginning to green up. In the second, the elongation stage, the distance between nodes gets longer, and the plant gets taller. This is the ideal stage for grazing because the leaves are well above the ground and the animals don't have to work very hard to obtain their nutrition. Depending on the species of grass, plants in this stage can be anywhere from a foot to several feet tall. In the third, the reproductive stage, the plant forms a flower head, and the energy shifts from leaf growth to seed production.

A grazing animal and a mechanical mower have very different effects on grass. An animal will chew the ends of the leaves, leaving the node intact so it can keep growing. A mower will chop everything down. After mowing, the grass must regenerate nodes from the crown in order to grow more leaves. If a plant has entered the reproductive stage, then it may not have enough stored energy to restart leaf production. In fact, mowing during the reproductive stage is a good way to eliminate undesirable grasses and weeds from a pasture.

It is amazing that any animal can not only live solely on grass but, in the case of cattle, grow to an impressive weight. It is all thanks to the magic of the ruminant's digestive system. The energy contained within grass is locked up in cellulose. Monogastric animals like humans and dogs have single-compartment stomachs that secrete hydrochloric acid, which can break down many substances, but they don't have the digestive enzymes that can further break down cellulose. Ruminants, however, evolved four chambers in their stomachs to digest cellulose. The first, and largest, is the rumen, which contains a soup of microorganisms that ferments the ingested grass. A side effect of this process is the generation of heat, which helps keep the animal warm in winter but can contribute to overheating in the summer. Another byproduct of fermentation is the production of methane and carbon dioxide, which go back into the atmosphere when the cow burps. (Contrary to popular belief, it is mostly the burping, not the farting, that is responsible for cattle's production of atmospheric methane).

Some grasses go through the three stages of development in the summer. These are the warm-season grasses, like bermudagrass. The cool-season grasses, like fescue, bluegrass, and ryegrass, grow primarily in the spring and fall. The holy grail of grazing is a mixture of warm- and cool-season grasses so there is forage available year-round.

The cows arrived at our farm in the middle of June, when the big pasture had been knee-high with bermudagrass. They'd grazed to their hearts' and bellies' content. I just had to make sure they had fresh water each day. It takes a lot of water to digest all that fiber. The rule of thumb is two gallons per one hundred pounds of body weight in the summer. For a one-thousand-pound cow, that's about twenty gallons per day.[2] Too bad bermudagrass goes dormant in the winter. In middle Georgia, that process begins around the beginning of October. Without any grass to eat, the cows would need hay for the winter.

Making hay is a practice as old as agriculture. The goal is to capture the nutrients in grass in a form that can be stored throughout winter. The steps in haymaking are simple: Cut the grass, ideally at the leafy stage. Let it dry in the sun. Rake into rows, called *windrows*. Bale it. In a good year, a haymaker might get three, or even four, cuttings out of a field. Many say that the second cutting is the best. The process, though, is labor intensive. For centuries, the scythe was the only tool for making hay. The swish-swish of the blade slicing through grass in the morning, when the dew makes cutting easier, had been part of the rhythm of farming for millennia. Now machines do the work. A tractor pulling a sickle-bar mower cuts the grass. A tedder fluffs it up. The hay rake piles it in windrows. And the baler gathers it up into neat packages.

Even with the proper equipment, successful haymaking is largely a matter of timing. You want to cut at the peak of leafiness. But to capture those nutrients, you need enough rain to get the growth. And if a field has been hayed for a long time, then the nutrients will have been depleted, so the farmer will have had to apply fertilizer at the right time: too early, and the chemicals leach out of the soil; too late, and the fertilizer does no good and money is wasted. Even when everything is done right, the haymaker is at the mercy of the weather. You need the rain for growth, but you need a window of several days of hot, dry weather for the drying process.

Making hay was beyond my capability. There wasn't enough pasture to set aside for haymaking, and I didn't have any of the necessary equipment. A used baler could easily fetch thirty thousand dollars at auction. A new one was double that. It was far cheaper, in terms of both money and labor, to simply buy hay from someone locally. Large, round bales would have been the most economical, but I had no way to move them, so I had to make do with square bales that could be lifted by one person. Again, Craigslist came to the rescue, and I procured one hundred square bales. With the hay stored up in the barn, I turned my attention to planting the pastures for winter grazing.

Planting a pasture the traditional way is labor intensive. First, the farmer turns up the soil with either a plow or a harrow. Next, he puts down seed. The seed can be broadcast with an implement that flings the seed in a semicircle behind the tractor, or it can be "drilled." A seed drill consists of a disc that opens up a furrow in the soil and a mechanism that drops the seed down a tube into the freshly opened ground. Finally, a heavy cylinder, called a *cultipacker*, is drawn over the field, tamping down the soil. Good seed-to-soil contact is key. You don't want any air pockets or the seed won't germinate properly.

Traditional seeding is a violent process that destroys the soil structure and makes the topsoil prone to erosion. A better way sows the seed with minimal soil disturbance. The implement of choice is a "no-till drill," an all-in-one solution that does the job of a harrow, drill, and cultipacker but without the magnitude of soil distur-bance that tillage causes. The no-till drill consists of three rows. The first row contains a series of discs (like a harrow at zero angle) to slice through dead grass and thatch. The second row has pairs of discs angled in a V to open up the slit created by the first row. Tubes connected to a seed box drop seeds into the V at a metered rate. The third row consists of rubber discs held under pressure by springs. These close up the soil, ensuring good seed-to-soil contact.

No-till drill. Seed is "drilled" directly into existing thatch without tilling. (Gregory Berns)

A no-till drill is the tool pre-ferred by farmers interested in sustainable agriculture, but they are expensive, complicated pieces of machinery. The local soil and water conservation district had one for rent, but their implement was so big I would have needed a tractor with hydraulics to lift it. My John Deere did not have rear hydraulics. Ken's did, but there was another problem.

Timing.

There was a limited window to seed the pastures. Plant too early and the cool-season grasses would germinate, only to wither in the early fall heat. Plant too late and they might not germinate at all. The best time for

  Cowpuppy

planting in middle Georgia ran from mid-September to early October. But if it rained, the pastures would become too muddy for a tractor. I could reserve the district's no-till drill a month out, but it would be a crapshoot to pick the right day. Everyone was in the same predicament, needing to plant when the conditions were just so. It was becoming clear why farmers had barnfuls of equipment. When something needed to be done, you couldn't wait around for someone to deliver what you needed.

My other neighbor, Wallace, had a no-till drill, which he used for seeding food plots. I hated to impose, but the alternative was to do nothing and hope for the best. He was gracious about it. When he was done planting his food plots, he could tractor over the no-till drill and seed my pastures. The timing would depend on the weather, so I had to be ready to go at a moment's notice.

The only thing that remained was to procure the seed. What to plant?

Middle Georgia is blessed with mild winters. Snow is almost unheard of, and while nighttime temperatures might dip below freezing, daytime highs are typically around fifty degrees. Several cool-season grasses do well under these conditions. Ryegrass and fescue are the two most common. Fescue is a perennial, meaning it comes back every year. The problem is that all the fescue in Georgia has been infected with a fungus that produces a chemical toxic to cattle when eaten in high quantities. Ryegrass doesn't have that problem, but it is an annual that has to be replanted every year. It is also favored for its fast growth.

Type of grass wasn't the only consideration. After years of being trodden by horses, the soil in the pastures had been compacted, which made it difficult to grow good-quality grass. When it rained, the water tended to run off rather than infiltrate the soil. Of course, I could have tilled up the field, but that would have been a temporary measure at best. A more sustainable solution used root crops to penetrate the hardpan. Daikon radish, known for its huge taproots, was ideal, and it grew well during cool weather.

Then there was the matter of fertilizer. It was expensive. And the synthesis of fertilizer burns fossil fuel. But plants need nitrogen to grow. Even though nitrogen comprises the majority of the atmosphere, grass can't just suck it out of the air. The only class of plants that can use atmospheric nitrogen is the legume family, which includes soybeans, peanuts, and clover. Soybeans and peanuts require warm weather, but clover is a cool-season plant, so that, I thought, should be part of the mix.

Ryegrass. Daikon. Clover. That was going to be the ticket to lush winter pastures.

I called the local seed-and-feed to place my order. Under better circumstances, I would have gone in person, but I had developed a love-hate relationship with the establishment, and phoning offered a degree of anonymity that increased the chances that I would be able to obtain the necessary supplies.

The seed-and-feed was ten miles away, in a run-down part of town. Shortly after moving to Tara, I had headed in to buy supplies for starting vegetables and to forge a relationship with the people on whom I would be depending. I strode in, wearing a KN95 face mask. Two ladies, perched on stools behind the counter, lorded over a store crammed with everything a farmer might need. Fifty-pound bags of deer-plot feed were stacked by the door. One aisle was devoted to equine supplies, and another to livestock. A side room was chock-full of gardening supplies. An assortment of rakes, shovels, and hoes hung from the wall.

The elder lady looked to be as old as the store itself, while the younger and friendlier one seemed closer to my age. Neither paid me any attention and continued chatting with the only other customer.

"Excuse me," I said. "Do you have any soil for starting seeds?"

Without looking at me, the older lady replied tersely, "You mean potting soil?"

"Mmm, no. I'm fixin' to start seeds indoors."

The other leaned over to the first and said, as if I wasn't there, "I think he means Pro-Mix."

The first jerked her thumb and said, "It's outside."

I went back out. But after two circuits of the parking lot, I couldn't locate what they were talking about. Reluctantly, I returned and professed my inability to find it.

Visibly perturbed, the old lady barked into a microphone, "Will, come to the front desk."

A few moments later, a rail-thin man materialized next to the woman. A Glock jutted prominently from his right hip.

The lady said, "He needs a bale of Pro-Mix."

Will frowned. "What's that?"

The lady lost all pretense of containing her disdain. She slid off her perch, mumbling, "Do I have a do everythin' 'round here?"

Will and I followed her outside. She jabbed a bony finger at a bale of soil wrapped tightly in white plastic and stomped back inside. It was, indeed, in plain sight. Will dutifully put it in the back of my SUV while I paid as quickly as I could to get out of there.

Maybe the seed-and-feed crew were temperamentally crabby, but I couldn't shake the feeling that I had been pegged as a mask-wearing, self-righteous city slicker. Even though I would have preferred to patronize local businesses, the experience had been unpleasant enough that I took pains to avoid going there again. This arrangement worked fine until I needed to procure seeds the big-box stores didn't stock.

Ken had told me that the seed-and-feed was under new management, so I called to see if they had any ryegrass seed. A man—the manager—answered the phone cheerfully. Promising as this seemed, the seed was on order.

"Should be in any day," he said.

I called back a few days later, but still no seed.

By the third try, the manager knew who I was. "Tell you what," he said. "Give me your number and I'll call you when it comes in."

After a week and no call, a decision had to be made. Either I kept waiting for the seed to arrive or I found another source. I am, as a general rule, risk averse. I think in terms of worst-case scenarios. The outcome to be avoided at all costs was missing the planting window because of an inability to procure seed. It didn't take long to find a supplier in Florida. The shipping would double the cost, but as a firm believer in avoiding regret, it was worth it.

On a crisp, cloudless morning in late September, Wallace tractored over with his no-till drill. Nine-feet wide, it could plant a dozen rows in one pass. Each seed had to be planted at a particular rate. Ryegrass would go at fifty pounds per acre; daikon at ten. Clover needed only three pounds per acre. We filled the main seed box with ryegrass and daikon. A secondary seed box, which metered out at a slower rate, was stocked with the clover. After a test run to check the seeding depth, Wallace began circuits of the big pasture. He planted nice straight rows, using the rear hydraulics to lift the drill at each end so as to make the turns. It took three hours to plant a total of eight acres across three of the pastures.

The timing couldn't have been better. It rained a week later. By the end of October, the ryegrass and daikon were coming up nicely. It wasn't enough to fully support the cows, but the hay I had stockpiled filled in the nutritional gaps. Each week, I moved the herd to a different pasture to give the grass some recovery time. Some graziers use more intensive management, moving the herd every day. Once a week struck a balance between what was best for the grass and the amount of time I wanted to devote to it.

Ryegrass is quick to establish, but it doesn't hit its growth peak until early spring. Just about the time my hay supplies were running low, the ryegrass took off, seeming to go from a couple of inches of stubble to lush pastures of knee-high grass. The cows couldn't keep up with it. Pretty soon I was back to mowing, finishing the job that the cows were supposed to be doing.

Because the no-till method had worked so well, I decided to invest in my own implement. It wasn't as big as Wallace's, but it got the job done. Every year, toward the end of August, I assessed where the cows had grazed the most and how well the previous year's plantings had done. With this information in hand, I placed my seed order for fall planting. Ryegrass was always in the mix, but I began to experiment with other grasses and grains, each year getting closer to the holy grail of year-round grazing.

October through December were the most difficult months. The summer grasses would go dormant, and the winter grasses would just be getting a foot-hold. Maybe it was due to climate change, but September and October seemed to be hotter and drier than I remembered them being in years past.

I learned about stockpiling to fill in the gap in fall forage. The technique was simple. In late August, I mowed a section of the big pasture as low as I could set the Bush Hog. Then, just before rain was forecast, I applied fifty pounds of nitrogen fertilizer per acre. Nitrogen is like steroids for bermuda-grass. The rain soaks it into the ground and the grass goes through a growth spurt for the next month. As long as I kept the cows off of it, the grass would regenerate. By October I could return the cows to that part of the pasture. The stockpiled grass would stay green until the first frost, which was usually in November. In a good year, the cows might need hay for only two months. I was beginning to feel like I was finally getting the hang of all this.

CHAPTER 7

Cow Love

If having a soul means being able to feel
love and loyalty and gratitude, then animals
are better off than a lot of humans.

–James Herriot, *All Creatures Great and Small* (1972)

As I grew closer to the cows and they began to accept me into the herd, their natural behaviors—both with one another and eventually with me—unfolded. It was in these quiet moments, me observing them and them observing me, that I became convinced of their capacity for love. I do not use the L-word lightly, as in, "I love chocolate." No, the cows evinced a depth of emotion and loyalty that surpassed even that of my dogs. It was all in their body language.

Although cows are by no means mute, they tend to vocalize as only a last resort of communication, like when Lucy bellowed for Xena or when Ethel mooed for treats. Instead of vocalizing, cows communicate with their bodies. It was only through patient observation that I began to understand their language and how they expressed their emotions.

My foray into the emotional lives of cows began with Ricky Bobby.

Ever since the day the cows came home to the farm, I had continued the daily ritual begun by the old man of feeding them a bit of grain every evening. He left me the red Folger's coffee can he used. So Ricky Bobby, Lucy, and Ethel already knew that when I brought out the can, treats would follow. Never mind that cows can't see red. The shape of the can and the sound of rattling grain were enough to get their attention. At first I spread the grain in a feeding trough, as I had seen the old man do. I soon realized that if I were to socialize them to humans, I needed a way to reward them for letting me touch them. Although I could feed them grain by hand, that was messy and imprecise. So in addition to the evening grain, I began feeding them cattle cubes.

One evening, after I had given each of the cows a few cubes in exchange for letting me scratch their heads, I mixed the remainder with the grain in the trough. This was the routine we had settled into. Ricky Bobby, true to his bull nature, lapped them up as fast as he could. It was a manifestation of how ruminants normally ate. His instinct was to gobble up as much forage as possible and then digest it later. Unfortunately, this was not a great strategy for cattle cubes.

Ricky Bobby gobbled up so many that a traffic jam lodged in his throat. He began backing away from the trough, his eyelids peeled back wide, showing the whites. Snot started running from his nose, and he shit himself in panic. Normal manure has a sweet smell, but this was sour and ran down his leg. The other cows didn't care. They just continued eating. More for them, I guess.

Choking is not a common occurrence in cattle, but it does sometimes happen when a momma cow attempts to eat the placenta after birthing her calf. It is not digestible for a ruminant, but they eat it anyway, probably to conceal the presence of the calf from predators. If it is big, the placenta can get stuck in the cow's throat. You're supposed to push it down with your hand.

There was no way I was going to stick my arm down poor Ricky Bobby's throat.

I approached him slowly, crouching so I could stroke his throat, thinking I could massage the blockage down. He recoiled a bit at the first touch. But I cooed soothing words to him while working the loose skin that began under his massive jawbone and flowed down to his brisket. I couldn't feel anything, but Ricky Bobby seemed to like it because he extended his neck a bit. After about five minutes of stroking, a deep belch reverberated through his gut, and I was rewarded with the smell of a sour burp in my face. And then he was fine.

"You're such a big baby," I said.

In the days following Ricky Bobby's big belch, his demeanor toward me changed. Instead of merely tolerating my entreaties to stroke his head, he began to seek them out. If I crouched down, Ricky Bobby would sidle up, waggling his head a bit, and then rest his massive noggin on my shoulder. If I didn't start stroking the underside of his neck, he would dig his chin into my back. The message was clear: *Scratch me.* The whole interaction was strikingly similar to what my dogs did when I rubbed their bellies. If I stopped, they would paw my hand, asking for more.

After a bit of rubbing his brisket, Ricky Bobby would respond by licking the nape of my neck. This was not a pleasant sensation. A cow's tongue is designed for tearing off bunches of grass with massive sharp-edged papillae. It is like a giant cat's tongue, except the cow uses more force. If I let him at it too long, he would lick me raw. Plus, I had seen Ricky Bobby use his tongue for things other than eating. As a full-blooded male of the species, he used it to sample the females' urine to see if they were in heat. Still, it was kind of adorable. I knew that his intention was kind, so I let him reciprocate by allowing at least a few licks.

Thanks to Ricky Bobby's gluttony, I had chanced upon a *calming signal* in cows. The term was popularized by Turid Rugaas, a Norwegian dog trainer.[1] Rugaas had observed that both wolves and dogs use specific behavioral signals primarily to avoid conflicts within their packs but also to calm themselves when they are stressed. For example, when strange dogs approach each other, they will usually signal a nonconfrontational attitude with relaxed postures and wagging, semi-erect tails. Conversely, if you approach a dog in a manner that makes them uncomfortable, it will turn its head to the side and often lick its nose. Yawning is another sign of stress. These are all signals to communicate that a dog doesn't want a fight. They also appear to have a self-calming effect.

All animals exhibit stereotypical affiliative behaviors when they want to appease another member of their species. Calming signals in dogs are just one example. Another is when cats purr. Chimpanzees groom each other. Elephants touch trunks. The behaviors are hardwired in each species, forming a sort of universal language within a social group.

Licking is the main affiliative behavior in cows. Sometimes a cow will ask another cow for a licking by putting its cheek near the other's mouth or gently nudging the other's nose or cheek. If she responds, she will lick the head and neck areas that are inaccessible to self-grooming.[2] Sometimes a cow will lick another without solicitation. In these situations, the cow might lick the back and rump in addition to the head and neck. Who licks whom is not random. Mutual grooming occurs more frequently between relatives, cows close in birth date, and subordinates on dominants.[3]

Why do cows lick each other? It could be that mutual licking serves a hygienic function by removing flies and parasites from areas that a cow can't reach on its own. If licking was purely hygienic, it would be more prevalent during the summer when parasites are more numerous. However, licking appears to be a regular part of a herd's daily life, independent of the season.

The limited research on cow licking consistently points to a social function. The outstretched neck of the receiver mimics the posture of a nursing calf, which is an example of *neotenous behavior.* Neoteny refers to the retention of juvenile traits in the adult of a species. Other examples include cats kneading and dogs licking their owners' faces. (Juvenile wolves lick their parents' mouths to induce them to regurgitate food.) These behaviors are hardwired from birth and never disappear. It is reasonable to assume that when an adult adopts a juvenile behavior it is comforting to them—a calming signal.

Licking serves the crucial function of maintaining the bonds between cattle. However, these bonds are neither random nor equally distributed in a herd. In the most extensive study of cattle relationships, researchers monitored a semiwild herd of full-sized African zebus over four years.[4] The scientists recorded which cows grazed near each other and who groomed whom. The relationships that emerged bore a striking similarity to those of wild chimpanzees and even humans. It was no surprise that mothers preferentially licked their offspring. What was surprising, though, was that this preference persisted into adulthood. Even after four years, a momma cow preferred to groom her offspring, and that preference was tilted to the firstborn, irrespective of sex. Licking also occurred between unrelated cows, but here, too, clear preferences emerged. Pairs of cows tended to graze near each other and lick each other, suggesting they formed stable friendships. And just like humans, some cows were more popular than others, getting lots of attention, while others nobody wanted to lick.

When I stroked Ricky Bobby's neck, I mimicked the effect of another cow licking him there, and he responded by relaxing and stretching his neck a bit. After that, he began actively soliciting me, just like he did with the other cows. Once the others saw what he was doing, they all wanted neck rubs. Except Lucy. It took six months before she let me scratch her behind the ears and then another six months before she let me work my way down her dewlap. But Ethel loved neck scratches and would always reciprocate with a sloppy kiss to my face. Xena was hit or miss, depending on her mood, but BB, the cowpuppy, would often start the process by licking me first.

It is tempting to ascribe these affiliative behaviors to the hormone oxytocin. All mammals, including humans, secrete oxytocin. And although it has been popularized as the "love hormone," oxytocin is responsible for a lot more than maternal-infant bonding. As a hormone, it flows throughout the body in the bloodstream, acting on almost every organ you can think of. Oxytocin also acts directly in the brain as a neurotransmitter, where it affects everything from memory to social cognition. Even men release oxytocin in small amounts.[5] Because of its wide-ranging functions, it is probably best to think of oxytocin as an amplifier, rather than determinant, of everything related to social cognition.

However, when researchers measured the concentration of oxytocin in the saliva of farm animals, they could not find a relationship to human interaction.[6] Even when the necks of cattle were stroked and the animals responded by stretching their necks farther, no consistent change in oxytocin was observed. This doesn't mean that oxytocin wasn't modulating the animals' reactions; it just means that the concentration in saliva wasn't the causative factor. The field of oxytocin research is rife with such negative findings. When it comes to behavior, it is the oxytocin in the brain—not the saliva, blood, or urine—that matters.

Oxytocin notwithstanding, I was not the first cattleman to interact with his herd in this way.[7] Although human relationships are less common with beef cattle because they are usually left on their own for days, dairy cows are a different story. The twice-a-day milking forges an intimate bond between cow and human, even if it is just to attach a milking device. It is an instinctive human behavior to stroke the cow while doing this. That cows also find it enjoyable was demonstrated in a 2008 study of how dairy cows respond to stroking of different parts of the body.[8] Scratching of the underside of the neck and withers—the area in front of the shoulder—elicited neck stretching and relaxation of the ears, while stroking the side of the chest did not. Rubbing the underside of the neck also lowered the cows' heart rates. The authors concluded that the cows perceived the human stroking similarly to social licking, if it was done in regions normally licked.

Emboldened by my breakthrough in bovine relationships, I perhaps overdid the neck scratching. In hindsight, this was a bit of a tactical error because as the previous research demonstrated, it was usually subordinate cows that initiated lick fests. So although I had succeeded in embedding myself into the cow herd, it is likely that Ricky Bobby came to view me as his subordinate. This was a potentially dangerous inversion of our relationship. Bulls are the most dangerous domesticated animal.[9] Anyone who works around cattle will tell you to never turn your back on a bull.

Pastor Ken thought my relationship with Ricky Bobby the most remarkable thing, but he also reminded me of what happened to his cousin. "He had a big ol' two-thousand-pound Angus," he told me one day. "I don't know what he did to that bull. Maybe he was abused. Or maybe he was jus' born that way. But I could see the meanness in his eyes. I told my cousin to be careful 'round that bull. But he didn't listen. Took a horn through the pericardium."

"What happened to him?" I asked.

"He died."

In cow licking, timing is everything. The herd engaged in social licking only when they weren't doing something else, namely their two favorite activities of eating and ruminating. In that regard, they were like dogs. Try to get close to a dog while they're eating and you're likely to get bitten. If I moved close to the cows when they were grazing, they would just amble on to a different patch of grass. If I tried to scratch them at the hayrack, they might shift position or tolerate my touching with barely disguised irritation. A slight narrowing of the eyelids made it clear they were saying: *Can't you see we're eating?* Ricky Bobby, though, made no attempt to hide his contempt by wagging his head at me, which was a warning sign to back off. On more than one occasion, I failed to recognize this signal and he turned to face me, head down, ready to bunt me out of his space.

Ricky Bobby had two types of head wags. The first was his relaxed, playful wag. If he was polite and lifted his head, I scratched his neck. Head up was a relatively safe position, and I tried to reward that as much as possible. The other wag warned of potential aggression. The mean wag had head low to the ground, chin tucked in, and horns pointed forward. It was a subtle but important difference.

The best time to insert oneself into the social structure was in the evening, after the cows had had their fill of grass or hay but before they settled down to chew their cud. I had started the evening checks after BB got his head stuck in the hayrack, and it was during these checks that the cows included me in their bonding rituals. Ethel was the most prolific licker. She directed the majority of her attention to BB. Then came Ricky Bobby. Sometimes Lucy would join in, and although she would receive licks from Ethel, she rarely reciprocated, reserving her licking for Ricky Bobby, who almost never reciprocated with anyone. I would bend down and rest my forehead against Ethel's, her horns cradling my ears. If I scratched her neck, she licked mine. Ricky Bobby inevitably would sidle over, wagging his head.

Getting the adults to accept me into their rituals was a process drawn out over the first year. Eventually, Lucy came around, but it took twelve months before she fully relaxed around me. What I had initially interpreted as a sort of aloofness was probably fearful uncertainty.

Xena and BB, having spent every day of their lives on the farm, required no acclimation to my scratchings and proddings. From day one, I was simply part of the herd, just like them. Xena had a streak of suspiciousness, but BB was everybody's friend. He was always the first to greet strangers, and he was the first to roll over for belly rubs.

Although BB eventually came to be the lead seeker of belly rubs, it was Ricky Bobby who started the routine, just as he had with the neck scratching. In late fall of the first year, I began sitting down with the herd during their midday ruminations. Unlike the evening sessions, there wasn't a lot of licking going on. From about one o'clock to three o'clock in the afternoon, the herd would pick a spot in the pasture to lie down in and chew their cud. The calves would snug up against their mommas, who fanned out from Ricky Bobby, usually rump-to-rump or back-to-back, depending on how cold it was. When the wind blew, they formed a tight scrum. If it was sunny and warm, they assumed a looser grouping.

One fine December day, I approached slowly, so as not to startle them, and sat down next to Ricky Bobby. He didn't flinch. He just let me plop down like the others. When I started scratching his neck, instead of sticking his head out farther, the big bull put his head in my lap and tipped on his side. All four legs jutted sideways, hooves pointing in the air. His massive belly heaved up from the ground in a great semicircle, his balls flopping against his thigh. It was a most vulnerable position, so I treated it with respect and stroked his belly gently. Ricky Bobby closed his eyes in extreme relaxation.

Ricky Bobby about to get a belly rub.

(Kenneth Peek)

I was acutely aware that even though the other cows feigned disinterest, they were paying close attention to the interaction. Their field of vision was almost 360 degrees, so they didn't have to look at me to see what was happening. Not surprisingly, BB was the first to pick up on this new type of social grooming. (I don't know what else to call it.) Within a month, all I had to do was approach BB while he was lying down and he would roll over.

Cow tipping is, of course, an urban myth. But cows do sleep on their sides. They just have to feel safe and be very relaxed to assume such a vulnerable position. Unlike neck scratching, though, there is no cow-cow equivalent of belly rubs. I never saw one of the cows roll on its side and have another lick it. So why did they start doing it with me?

There were several potential explanations. It could have been another neotenous behavior. Momma cows frequently groom their calves, paying particular attention to the genitals. Scratching the belly might make the cow feel like she did when her momma groomed her. Or maybe it just felt good. Cows can't reach their own undersides easily, so maybe they appreciated the sensation. Or it could have been another calming signal as a way of showing nonaggression to each other. Likely, belly rubs were a combination of all three.

Between the neck scratches and the belly rubs, the emotional bonds between the cows and with me were obvious to anyone who visited the farm. Ken, of course, saw all of this evolving from the day Lucy, Ethel, and Ricky Bobby came home. None of this surprised him, except for a bull putting his head in my lap and rolling over. But Ken knew well that all of God's creatures had emotional lives. It just took patience and a willingness to communicate with them on their terms to connect with them. Even the question of love did not seem at all strange.

In moments when I questioned the value of what I was doing on the farm, Ken reassured me. "I've been around a lot of cows, and I can tell these appreciate you. They're so comfortable, they roll over for you. That's something special."

There *was* something special about it. But was it just a common behavior or was I potentially discovering a new breakthrough of some sort? I was a scientist, after all, and this whole journey seemed to be turning into something more than just a new farmer getting to know his cattle. Maybe it was like the rush a lion tamer feels when he sticks his head in the mouth of an animal that can kill him. But I think it had more to do with the depth of connection. The cows were smart. Emotional. And they could hold grudges. They had a more human range of emotions than dogs. The momma cows stayed emotionally attached to their calves for life. How must they feel when they're separated? And unlike the bonds I had with my dogs, the cows' love was hard-earned. Strong herding instincts meant they were wired to stick with their own kind. There was a thrill about being let into their world. An honor, really.

CHAPTER 8

Dogs Versus Cows

If you tire, give me both burdens, and rest
the chuff of your hand on my hip,
And in due time you shall repay
the same service to me,
For after we start we never lie by again.

–Walt Whitman, "Song of Myself," *Leaves of Grass* (1855)

By early March of the second year, the remnants of winter were fighting a losing battle against better weather.
Winter's last tendrils manifested mostly as morning frosts, but
the days belonged to spring. I looked forward to April and May,
one of the two times of year when the atmosphere in the South
achieves a delicate balance of warmth without humidity. The
other is spring's counterpart in early October.

The pastures were just beginning their vegetative growth
spurt, and although they were noticeably greener every week, the
grass wasn't thick enough to fully support the cows' nutritional
needs. I still gave them a few flakes of hay every day along with a
bucket of corn and sweet feed. As I drove home, two fifty-pound
bags of corn weighing down the back of the car, the landscape
sparkled in the sunshine. The dogwoods were beginning to
pop, their white flowers peeking through the branches like
little origami fortune tellers.

A mile from home, three cars were pulled over on the shoulder. This was not a normal occurrence. A trio of women scuttled about in a frenzied state. I rolled down my window and asked if they needed help. One woman, who seemed to be the most agitated of the bunch, fluttered her hands and wailed, "There's a dog running back and forth across the road. She's gonna get hit!"

How could I not help? Dogs, after all, were my kryptonite.

The stray in question was a medium-sized white dog with black button ears. As soon as I got out of the car she ran to me, wagging her entire rear in happiness, as if to say, *Where have you been all my life?* This was all the more remarkable because apart from her obviously sweet disposition, her most distinguishing feature was the baseball-size tumor jiggling from her right front leg. I squatted to get a closer look and was rewarded with a slobbery kiss that smelled like fish. Of course she had no collar. Most of the dogs in these parts ran free without tags. Her nails curved in inch-long arcs, forcing her toes to spread in an undoglike manner. None of this seemed to bother her, though. She was just happy to be the center of attention.

The frantic woman asked me if I lived around here, which of course I did, and if I knew the dog. The woman wore a tight-fitting black leotard top. Her hair was nearly as black and flew back and forth in loose waves around her face, as she seemed to be in constant motion. Another woman, who came from a different car, said she lived in Brooks, which was five miles farther down the road and across the river. The third walked up from a driveway across the road and informed the crowd that the homeowners didn't recognize the dog.

The woman in black blurted, "Someone dumped her!"

The Brooks woman agreed. "It's true. People from the city are always dumping their dogs out here. They think some kindly farmer will just take them in."

I wasn't so sure. Apart from the toenails and the goomba on her leg, the dog was in good health. Her dense fur was soft and clean, whereas strays usually got covered in mud as they thrashed about trying to find a way home. But it was clear that nobody was going to do anything for the dog. The women's eyes all said the same thing, praying that someone else would take responsibility for this creature.

The dog *was* sweet. I crouched down and picked off a tick from her eyelid. Her eyes beamed pure trust.

I sighed and said I would take her.

I thought the woman in black was going to cry, and I accepted her offer of a hug. When we unclasped, she said breathlessly, "Thank you. Thank you. There is a special place in heaven for you."

I didn't know what to say, so I just lifted the dog into the car. She headed right to the front passenger seat as if she had been doing this all her life.

We drove the remaining mile, just man and dog, like we were meant to be. The women followed me to the driveway and waved goodbye as I pulled in. I never saw them again. As I drove up to the house, I wondered whether the woman in black might have been the one who dumped the dog, but I pushed that thought out of my mind. Whatever the cause, the dog happily riding shotgun was about to become part of Talking Dogs Farm.

Bringing a new dog into the house was not something I took lightly. The four current canine residents were all over ten years old and had had many years to work out their social structure. None of them were friends with one another, but they coexisted peacefully. Thor was a big German shepherd mix and the leader of the pack. We had inherited him from Kathleen's father in California, who had suffered from dementia. When Thor became too much for him, we flew the dog to Atlanta with the intent of rehoming him. He lasted two weeks with his new family before they sent him back to us. Apparently Thor did not take kindly to their eight-year-old boy hugging him. Thor was not subtle. He made his displeasure plainly known by a growl emanating from deep inside his chest. Failure to heed the warning would result in the baring of teeth and escalation to growling that, upon both inspiration and expiration, sounded like the devil's accordion. Yet Thor had never bitten anyone, most likely because his verbal warnings were so frightening. No vet would examine him without a muzzle, and the other dogs in the house had no choice but to respect him. If Thor accepted the new dog, so would the others.

Thor greeted the visitor with characteristic enthusiasm. After the obligatory sniffing of butts, the two dogs ran around the backyard playfully. Thor asserted his dominance by trying to mount the new dog, but she deftly elided his display by wriggling away. Apparently this little dog's charm wasn't restricted to people. Never had I seen Thor accept a strange dog so quickly—or ever, for that matter.

With Thor's stamp of approval, the three other dogs followed suit. Callie, the little black terrier with whom I started the dog MRI project and whom I adored the most, had a history of picking fights with strange dogs, but even she seemed to like this newcomer. Callie started doing doughnuts, running tight circles around the dog. Cato, a constitutionally anxious Plott hound, hung back and barked at the ruckus, while Argo, a generic yellow dog, tried to nip the new dog's heels. Argo took the longest to warm up, but eventually he came around to accept her.

Because of her easygoing personality, we named her Ripple, after the song by the Grateful Dead that channeled Walt Whitman's sentiment about companions for life. Certainly not the fortified wine of yore, as most visitors to the farm assumed. No matter. She didn't have a mean bone in her body.

How was it that Ripple could instantly become everyone's friend, while the cows took months to get warmed up? Although BB had doglike traits, especially in his capacity for affection, calling him my cowpuppy was more a term of endearment than an accurate description of behavior. The fact that he sometimes acted like a puppy was the result of his nearly constant socialization with me from the day he was born rather than the biological predisposition that dogs had. Dogs were easy. Cows, not so much. Both, however, were capable of forming strong bonds with humans.

In terms of winning our hearts, dogs had a head start coevolving with humans before the wild aurochs—the giant mammal that was the ancestor of cattle—was tamed. By the time Mesopotamians domesticated cattle, dogs had been living in and around people for several thousand years. With dogs' rapid reproductive cycle, that was plenty of time for humans to select for traits they found desirable. When early cows came into existence, dogs were already enmeshed in human society. They probably even helped guard the first cows.

But why were dogs the first animal to be domesticated? After all, they evolved from wolves, who were not known for friendliness to humans. Several factors brought the two species together. It was the end of the last Ice Age, and glaciers covered much of Europe. This left an arc of temperate climate along the Mediterranean. Species of all sorts shared this land. Wolves and humans were the apex predators and competed for some of the same food. This similarity of diet would prove to be crucial to the emergence of the dog. A leading theory of dog domestication posits that the friendliest of the wolves began hanging around human encampments, scrounging off leftover food and garbage. Those that could tolerate being close to humans bred with each other, further establishing a line of tame wolves. These proto-dogs were used to traveling long distances and had no problem following humans wherever they went. When humans created semipermanent cities in the Fertile Crescent, the nomadic protodogs no longer needed to hunt on their own. They became fully dependent on humans by being fed either directly or indirectly from scrounging the growing garbage heaps.

Two factors, then, explained the origin of dogs: their ability to follow people during their migrations and their capacity to eat human food. The similarity of our diets was crucial to the formation of the dog-human bond. So much of human social life centers around food; it would have been impossible to ignore the plaintive stares of dogs hanging around the camp-fire. The dog and human digestive systems are similar enough that not only can a dog eat what a person eats, a human can eat what a dog eats. Dogs adapted even further to the human diet by evolving genes that allowed for the digestion of starch.[1] In contrast, cows' digestive systems are so different from humans' that there is very little in common between what both cows and people can eat. Although cows do have a sweet tooth and enjoy sugary human snacks, humans can't digest the mainstay of the cattle diet: grass or hay.

And because wolves were predators, dogs retained the confidence that went along with hunting. A predator looks at the world and asks: *Can I eat that creature?* Prey are always wondering: *Will that eat me?* Totally different worldviews and completely different psychologies. This predatory confidence allowed dogs to become the first species to shed their fear of humans. Given that, why do wolves keep a healthy distance from humans? Because modern wolves are the descendants of the wolves who decided to stay away from people. The ancient wolf doesn't exist anymore. The population split into the human-oriented dog and the human-fearing modern wolf. By hitching their wagon to humans, dogs conquered the world. There are approximately 800 million dogs in the world but only 250,000 wolves.

Although the general process of domestication of dogs and cattle was likely similar, the fundamental differences in their diets put a hard constraint on the types of relationships humans can have with them. Humans can sleep with dogs because they don't poop in the bed. That is a direct consequence of the slow digestion of a carnivore. Ruminants, like cows, are continually moving roughage through their digestive tracts and crap constantly. Plus, they only sleep three hours a day because they are wired to keep an eye out for predators.

These biological differences carry over to psychological ones. The biggest difference between dogs and cows I had seen was in their ability to understand human language. Dogs have an obvious ability to learn simple elements of human speech. Take names, for example: every dog in our house knew its name. They came when called (usually), or at least oriented to me. Even Ripple, who must have been called something else at some point, immediately learned her new name. But among the cows, only BB consistently oriented to his name, stopping what he was doing and turning his head to look at me. Often he would come to me. Perhaps that was why I was so drawn to him, or maybe it was because of that extra attention he received. None of the others showed such a response. Sometimes they would rotate an ear in my direction, but I couldn't determine if it was specific to their name or just the recognition of my voice. None of this is to say the cows didn't know their names. Maybe they did, but because I didn't have anything of value to them, they paid me no attention.

There was nothing in the cow brain that indicated they couldn't understand vocal commands. Their hearing was probably as good as dogs'. And cows' vocal range, which I would characterize as the equivalent of a human baritone, was similar enough to a person's that the cow should be able to distinguish between different human utterances. Most likely it was a training issue. The cows were always together as a herd, so it was difficult to call an individual without everyone hearing. If each of them had had more one-on-one time with me, speaking their name so only they could hear, then they might have shown more specific orientation responses.

There were also fundamental differences in their social architectures that made dogs more able to respond individually than cows. We think of dogs as pack animals and cattle as herd animals. The semantic difference is that *pack* describes a group of predators, whereas *herd* is a group of prey. Packs and herds behave differently. Prey animals herd together for protection. When under threat, the individual that stands out is at greater risk of being killed. So it made sense that when the cows grouped together in a herd they wouldn't respond individually to my calls. Dogs and their wild cousins formed packs, but their goal was not protection. It was to coordinate actions to achieve a common goal of bringing down prey. For wolves and coyotes, this might mean the individuals were separated by significant distances but still in pursuit of the same prey. So even though dogs were pack animals, they still maintained their individuality, making it easy to teach them their names. The cows probably knew their names, too, but they might respond only when the herd was comfortably dispersed and there was no pressure to lose their individuality.

What about learning commands? It's obvious dogs aim to please. As long as the human is consistent in what they ask a dog to do, dogs can learn an impressive range of tricks. The basic "sit" command is simple for a dog because it builds on a natural behavior: sitting plus looking into the eyes of the owner, waiting for a treat or some praise. When my colleagues and I used MRI to study dogs' motivations for following human commands, we found that the majority of dogs' reward centers were driven as much by the expectation of praise as by the prospect of a food treat.[2] But I had yet to teach one of the cows a single vocal command. I hadn't been able to get BB to lie down on command. This was mainly a result of my asking them to do unnatural things. As beasts of burden, cattle had been employed for millennia in many sorts of jobs, usually pulling things. If started young, cattle can easily be trained to wear a halter and, later, a harness for pulling. I had taught BB to wear a halter, but I hadn't taken it to the next step.

It may seem strange that it is harder to teach a cow to lie down than it is to teach her to pull a cart. But here's why. Whenever I crouched down to scratch a cow's brisket, they would lean into me. Ricky Bobby was the most exuberant in his leaning, but BB also acquired this tendency. There was nothing magical about it. All the cows had a reflexive action to oppose any force pushing on their body. If I rubbed Ethel's neck hard, she would turn her head the opposite way and press her neck into my hand. Likewise, the cows used their heads to push against people and things. So when a yoke is placed around a cow's neck, they already have a natural instinct to push against it. If a cart is tied to the yoke, then the cow will pull it along with them. All that remains is to teach them four commands: go forward ("come up"), stop ("whoa"), turn left ("haw"), and turn right ("gee").

Despite these differences in language comprehension, the cows showed a remarkable similarity to dogs in their demonstration of emotions. Charles Darwin had argued that there was an evolutionary continuity in the expression of emotions in both man and animals. There are, of course, the basic emotions like pleasure and fear. Every animal has those. But what endears dogs to us is their apparent capacity for what we take as love. If we accept that there are different types of love, then it is easy to recognize a dog's version. Some researchers have argued that a genetic mutation is responsible for dogs' innate ability to bond with humans, which bestowed the capacity to love us.[3] The love of a dog is not solely dependent on food, as the behaviorists of yore would have us believe. Love is nontransactional—it is not contingent on something else. You know your dog loves you if it just wants to be around you, even when you don't have food. I had had many dogs in my life, but only a few were like that. Callie—the first MRI dog—loved to do things with me, which is how I was able to train her to go in an MRI scanner.

And Ripple. I had had no history with her. She just walked into our lives and opened her heart to everyone. I wish I could have spent more time with her, but the tumor on her leg was just the tip of the iceberg. She was filled with cancer. Six months after we adopted her, one of her internal tumors burst, and she collapsed on the floor. It was blessedly quick. It was the only time I had had a dog leave us on their own terms, without the dreaded final trip to the vet.

Dogs were easy to love, especially ones like Ripple.

The cows took effort. But as I had discovered with BB, it was an immensely rewarding effort. Nobody knows if cows have the same mutation for love as dogs do, but I think it is unlikely. The bovine and human lifestyles are not nearly as compatible as that of dogs. So there hasn't been as much genetic pressure to select for companionship traits in cows. This doesn't mean, however, that they are incapable of love. It just takes more to win their hearts. But that makes it much more special when it does happen.

In addition to a dog's love of wanting to be around us, we can read their emotional states from body language and the obvious tail-wagging. It wasn't too different with the cows. Much of their body language was in their head carriage, which made sense because cows use the head for different things, whether for eating and drinking or as a battering ram. Interestingly, they also wagged their tails like dogs. The calves always did this when they were nursing. It was easy to spot because a nursing calf lifted its tail near vertical. The switch would dangle freely, and when the calf wagged its tail, the switch waved back and forth like a flag. The only time the adults wagged their tails was when they got the cow zoomies. When they were excited and chasing each other around, they held their tails high, letting their switches stream behind them in the wind.

In terms of cognitive and emotional skills, cows had much in common with dogs. Most of the differences could be attributed to the divergent worldviews of predators and prey. Because humans are predators, too, it is hard for us to understand how a prey animal, like a cow, sees the world. Analysis of their visual systems, though, reveals much about what it's like to be a cow.

CHAPTER 9

Through a Cow's Eyes

The eye is the lamp of the body.

—Matthew 6:22

It's inevitable that when you spend time with someone, you gain an appreciation for how they see the world. The hay-rack episode turned out to be a fortuitous accident. No harm came to BB, and with the extra attention the cows received in the evenings, I began to understand how they perceived their environment. This, in turn, allowed me to approach them in ways I wouldn't have thought of otherwise, and these sunset bonding sessions formed the basis for my relationships with them. In the same way, they began to see me in a different light.

Although cows have excellent hearing and a keen sense of smell, vision is their primary sense for interacting with the world, as it is for humans. But even though cow vision is similar to human vision, there are several differences that explain much of their behavior.

As prey animals, cows have a preternatural sense of motion. They are always on the lookout for predators, whether it is a coyote, a stray dog, or a human coming to do them harm. I was reminded of this sensitivity during many of our evening sessions. Dusk was a period of great activity around the farm. Not just the cows but all the other critters came out at night. The colony of bats living in the barn loft would begin their exodus shortly after sunset. They would flit out from beneath the roof peak as I filled the cows' grain trough. They emerged one or two at a time, zigzagging their way across a pink sky. The cows did not seem to pay them any attention, at least not that I could detect.

The deer, though, were a different story. Ethel was usually the first to alert to their presence. She would stare at the tree line on the edge of the pasture. The first time she did this, I didn't know what she was staring at. It was creepy, a cow just staring off into space, like something out of a horror movie. At first, I looked where she was looking and couldn't see anything. Only after repeatedly scanning the tree line could I make out the subtle movements of the local herd of deer. Their summer coats were chestnut brown, and in the evening light, they took on a hue almost identical to that of the red Georgia clay. If they didn't move, they were invisible—except to Ethel. She did not appreciate interlopers in her pasture. She snorted softly, a muffled version of the exhalation she used to express her displeasure with the other cows, like Marge Simpson quietly expressing disapproval.

BB, however, stayed true to his personality and showed no interest in what was happening in the forest. He was content to accept my offer of scratching his brisket while his momma stayed alert for distant threats. Lucy also kept an eye on the horizon, but because she was either wiser or more stoic than Ethel, it wasn't as obvious what she was doing. Plus, her dark brown eyes blended in with her black fur, making it harder to discern where she was looking.

It is an axiom of evolutionary biology that form follows function, and the configuration of the cow's visual system is such that it is perfectly adapted for the cow's natural environment. Close examination of the cow eye reveals much about its visual world and why Ethel behaved the way she did.

At a basic level, the cow eye is

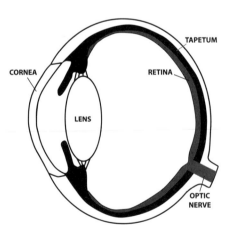

Anatomy of the cow eye.

no different than the human eye. It is a globe with the cornea at the front and the retina in the back. Light enters through the pupil, passes through the lens, and hits the retina, where it is transformed into nerve impulses that travel to the brain through the optic nerve. However, the cow eye differs in several important ways from the human eye, reflecting the different visual needs of the cow. Instead of a round pupil, the cow has a horizontal oval that can constrict to a slit

Ethel's eye, showing the slit-like pupil oriented horizontally. (Gregory Berns)

in bright light. Horizontal pupils are characteristic of all ruminants, in contrast to the slit-like pupils of predators, like cats, that are oriented vertically.[1] Horizontally elongated pupils let in the maximal amount of light parallel to the horizon, where predators might be lurking. The cow's retina mirrors this geometry. Whereas primates have a central fovea with cones packed tightly for maximal visual acuity, cows have a visual streak spread horizontally across the retina. This is why Ethel detected motion in dim light while I had difficulty. Her motion detectors were tuned to monitor the horizon. Also, cows, like dogs and cats, have a reflective layer behind the retina. The tapetum reflects back photons that slip between the photoreceptors, effectively doubling the light-gathering capacity of the eye. The tapetum causes the ghostly glow you see when shining a light in an animal's eyes at night. Finally, the cow's eye is astigmatic, meaning it is not round. It is as if they are wearing bifocals. They can focus on objects close to them if they are below them, on the ground, whereas the upper field of vision focuses on things far away—perfect for an animal that spends much of its time looking for things to eat on the ground while keeping an eye on the horizon.

The most obvious feature of cows' visual systems is the position of their eyes. All ruminants have eyes set in the side of the head. Cows' eyes are set fifty degrees from the long axis of the nose, whereas those of predators like dogs and cats are set ten to twenty degrees off-axis. (Human eyes are aimed twenty degrees off-axis, like dogs.) The rabbit, a highly preyed upon creature, has the most extreme positioning, with eyes angled eighty-five degrees from the midline. The wide-set eyes of prey animals reflect the evolutionary adaptation that expanded their field of view to see all around them.

Cows' expansive visual field comes at the cost of binocular vision—the area where they can fixate both eyes on an object. Having both eyes on something allows an animal to precisely gauge its distance. This is a critical skill for humans and other primates who use their hands to pick up things. The importance of binocular vision for depth perception can be demonstrated by closing one eye and touching a nearby object, like a coffee mug. It's much more difficult with only one eye.

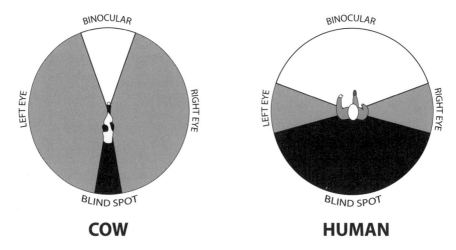

COW   **HUMAN**

Visual fields of cows and humans. The cow has an expansive field of view with a blind spot just behind the rump, but this comes at the cost of binocular vision in the front.

Grazing animals still need to judge distances. Depth perception may be less important for eating grass, which is mostly a tactile experience, but it is critical to avoid running into things. Even though cows can't lay both eyes on a lot of their visual field, they have workarounds for judging distances.

Cows can simulate binocular vision by simply moving their heads back and forth. The motion affords a single eye with different vantage points and is an effective substitute for having two eyes on an object. You can demonstrate this motion parallax for yourself by closing one eye and holding your hand in front of your face. It is hard to perceive depth, but if you shake your head left and right, the two-dimensional image magically becomes three-dimensional.

Xena was the most prolific head-shaker in the herd. The other cows would often push her aside at the feeding trough, and she would stand there looking forlorn, seemingly wondering why even her momma, Lucy, shoved her away. The maternal bonds apparently had limits, especially when it came to high-value food.

I was a sucker for Xena's baleful eyes and always offered her a handful of grain. An outstretched hand, full of sweet feed, would seem to be a no-brainer to a cow. But you have to consider it from the cow's point of view. Although they have binocular vision in front of them, their long muzzles result in another blind spot just in front of their noses. From Xena's perspective, my outstretched arm disappeared into this black hole. The only way she could decide whether my hand contained a treat was to waggle her head back and forth. This type of head waggle was much more rapid than the slow wag the cows used when they wanted something.

Then there is the question of color vision. For some reason, many people think that a lot of animals—dogs included—are color-blind, seeing the world in shades of gray. Dogs are not completely color-blind. They are dichromats, meaning they have two types of cones in their retinas, one responsive to blue light and the other to green. Cows are similar.

Most humans, in contrast, are trichromats. We have three types of cones, responsive to blue, green, and red light, respectively. However, about one in twelve males is born without one of these types of cones, usually the red one, resulting in difficulty discriminating between shades of red and green.[2] (A similar percentage of females may have a fourth type of cone, resulting in an enhanced ability to perceive colors.)

The perception of color, though, is a fair bit more complicated than simply stimulating different cones in the retina. The human brain takes the relative levels of cone activation and computes three different images in parallel.[3] First it constructs a luminance image by summing the signals from the three types of cones. This is the grayscale version of a scene. The brain also constructs two difference images: one as the red minus green cones, and another as the sum of the red and green cones minus the blue. These three images are analogous to the three-color channels of a color image displayed on a computer screen or in a printed book (in CMYK space—cyan-magenta-yellow-key, where key is analogous to luminosity and also referred to as black level). Someone lacking the red cone will not have an R-G channel and will therefore perceive the world in shades of blue and yellow.

As dichromats, cows have two types of cones. The peak sensitivities occur at 451 nanometers (blue light) and 555 nanometers (green light), but in actuality, the cones' sensitivity is spread over nearby wavelengths.[4] This is very similar to dogs' visual systems, except the sensitivity of the canine blue cone is shifted more to the violet end of the spectrum. Because cows lack a photoreceptor for red, they don't perceive red colors the same way humans do. So even though a toreador's cape is bright red, it is commonly believed that bulls are attracted more to the motion than the color.[5]

But the fact that cows lack a cone sensitive to red does not mean they can't tell the difference between a red object and something else. In one study, researchers trained a group of calves to select the brighter of two lights by giving them a food reward for correct responses.[6] One group of calves was trained using blue lights, one with green lights, and another with red lights. When they could perform this discrimination with 75 percent accuracy, the researchers substituted a different color for one of the lights. If, in fact, the cows were unable to tell the difference between, say, red and green, then their accuracy should drop to 50 percent. But that wasn't what happened. The calves trained on the red light were able to discriminate between both red and green and red and blue. Inexplicably, the calves trained on blue light performed at chance for both green and red. An older study of fighting bulls came to similar conclusions.[7]

So cows can't, technically, see red. Yet they can learn to tell the difference between red and other colors. How do they do that? The key is in the computations the brain performs. Like humans, the cow's brain computes a luminance image and a difference image. Except instead of two difference images, there is only one: green minus blue. Although it is imperfect, we can get an idea of what a cow sees by using software that simulates what a human lacking the red cone sees.[8] An image of an apple, for example, appears more monochromatic than other colors but is still differentiated easily from other colors, although not from other colorless objects like Ethel's fur.

The color limitations are irrelevant to cows. If seeing red was important to a ruminant, evolution would have selected for that ability. For color vision, what matters to the cow is finding the best forage to eat. Everything else is aimed toward predator detection, namely, an expansive field of view, excellent night vision, and superb motion sensitivity.

These visual adaptations, while serving the cows well in terms of monitoring for threats, were also the main impediments to forming close bonds with them. Our evening sessions made clear how they saw the world, but to use this knowledge to get close to them required a deeper understanding of how these visual processes formed the boundaries of their personal space.

CHAPTER 10

Personal Space

This constant watch for enemies . . . is the wild animal's chief preoccupation. It is ever on the alert, so as to avoid enemies and be ready for escape. This perpetual never-ending activity, even during sleep, represents so fundamental an occupation–overriding all other behaviour.

–Heini Hediger, *Studies of the Psychology and Behaviour of Captive Animals in Zoos and Circuses* (1955)

With the knowledge you've been given, you are now on the inside of what I like to call the Byrnes family circle of trust.

–Robert De Niro as Jack Byrnes, *Meet the Parents* (2000)

When the old man told me the bull was friendly, I didn't know if he meant the females were unfriendly or simply indifferent to humans or that Ricky Bobby possessed exceptional comity. In any case, the man appeared to have a conventional relationship with the cows, sticking to the basic feeding requirements and not cozying up to the herd as I was doing. In a demonstration of mutual respect, he had used his walking stick to keep the herd at a safe distance. He stayed out of their space and they stayed out of his.

89

When my evening sessions with the cows became part of the daily routine, Ricky Bobby remained true to his character. BB and Xena were easy to approach and pet, but they were calves and didn't know any better. It was Lucy and Ethel who showed the most intriguing developments in behavior.

I was determined to get up close and personal with all the cows because that was the only way to learn about their world. Ricky Bobby made it easy. The impressive reach of his tongue allowed Ricky Bobby to extend his personal space anywhere his tongue could reach. This included blades of grass, treats, the feed trough, and any orifice emitting interesting odors. He had no issue with me scratching his head as long as I kept supplying the tongue with a steady stream of cattle cubes. His friendliness, though, was not without bound. Sudden movements, unfamiliar objects, and loud noises would trigger his flight response, and he'd be the first to hightail it to safety.

For some time, the females did not share Ricky Bobby's lack of boundaries. Ethel didn't let me scratch her head for a month. Lucy, though, was the most difficult nut to crack. She didn't loosen up until well into the fall of the first year, and even then it was for only the briefest of chin scratches. But by the second year she was the most loyally affectionate of the bunch. It almost made me cry when she sidled up to me, gently rubbing her horns against my leg—the cow equivalent of a dog resting his nose in your lap. How we got to this point in our relationship revealed more about the personal space of cows than anything else they did.

The idea of personal space—a sort of bubble surrounding a person or animal—is a psychological concept that began with the study of animals in the 1950s and crossed over to the popular lexicon to encompass humans. Heini Hediger, a Swiss biologist, pioneered the study of wild animals in captivity, a field that came to be known as "zoo biology." Hediger observed that animals in their natural habitats were in a constant state of vigilance. He concluded that the chief concern of every animal was to avoid being killed. Everything else—including food and reproduction—was secondary. After all, a dead animal can't reproduce.

So as not to be killed, all animals surround themselves with an invisible zone. Incursions into this zone are potential threats. Hediger called it the *flight zone* because entry into the zone triggers the animal's flight response. Some domestic animals, notably dogs and cats, have lost much of this response, to the point that their flight zones are sometimes nonexistent. But dogs and cats are predators, with few enemies of their own. Domesticated prey animals, like cattle, horses, sheep, and goats, retain much of the flight instincts of their wild ancestors, albeit in reduced forms. Their predilection to flee anything perceived as a potential threat makes getting close to them a challenge—and sometimes dangerous.

There are at least two ways to talk about an animal's space.[1] In the geographical sense, space equates to territory. For many predators and the prey they hunt, this kind of space can range for miles. A wolf's territory, for example, encompasses two hundred to four hundred square miles.[2] In the second sense, space refers to an animal's flight zone. This is the distance an animal places between itself and others. Different species have different characteristic flight distances, although this can vary among individual members of a species. Apart from size, the main difference between personal space and territory is that the animal carries around its personal space wherever it goes, whereas territory is fixed. Also, because flight distance isn't static, an animal's personal space expands and contracts as the circumstances dictate. Good animal handling creates an environment in which that bubble shrinks, allowing the handler to enter an animal's personal space.

As researchers studied flight zones, it became clear that personal space is composed of at least three separate zones arranged in onion-like layers around the animals. Hediger's notion of a flight zone is actually the outermost layer. This zone is called the *extrapersonal space* (EPS) and potentially extends as far as an animal can sense. Anything within sight or earshot is probably in an animal's EPS. Odors, too, can travel long distances to reach an animal's nose, placing the source within the receiver's EPS. The middle layer—the *peripersonal space* (PPS)—is the zone immediately surrounding the body. The PPS encompasses the area within an animal's reach, or, for those without hands, the space in which things

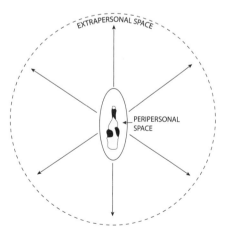

Personal space of the cow. The peripersonal space is the zone of direct actionability. Extrapersonal space extends as far as the cow can see.

can be acted upon directly.[3] Although originally conceived as a buffer zone against threats, the PPS also represents the space in which an animal can immediately acquire things of utility, especially food, but also social rewards like the comfort of herd mates. The final layer—the skin itself—is the last line of defense. It is the physical demarcation between an animal and its environment. The skin is loaded with neural sensors that provide the brain with a constant stream of data about the state of the environment. Unwelcome incursions in this layer result in a slew of immediate reflexive responses: skin rippling, tail swatting, kicking, biting, and running.

To work with a preternaturally flighty species like cattle, a human has to gain access through all of these layers. First, you have to penetrate the extrapersonal space—the flight zone—without the cow running away. Then, if you get close enough and you want to touch the cow, whether for pleasure, milking, or the occasional veterinary procedure, you have to enter the peripersonal space. Finally, you must breach the personal space of the skin itself without the animal injuring you or running away. As you might imagine, gaining access to each of these zones requires increasing levels of trust. There are no shortcuts, and you can't bypass any layer.

When Ricky Bobby, Lucy, and Ethel—the OGs—arrived on the farm, I mistakenly wrote off Lucy. She had little interest in people other than as providers of food. Sure, she lapped up the sweet feed I put in the trough and even tolerated my presence while eating it, but as soon as the grain was gone, she was the first to head back to the pasture. Ricky Bobby would hang around, thrusting his tongue in the air, begging for more, but Lucy would not debase herself with such behavior. I figured that was just the way she was.

I had no doubt the cows were closely monitoring my movements in their extrapersonal space. I had already seen evidence of their acute visual perception, especially for movement in dim light. Just as when the deer encroached from the tree line and the cows oriented to them, so they did when someone left the house. I first noticed this when working at the barn and saw both Lucy and Ethel turn to look at the house. It took me a while to see that it was Kathleen on the front porch. The barn and the house were separated by one thousand feet, indicating that the cows' EPS extended at least a quarter mile from their bodies.

When something enters an animal's EPS, two processes occur simultaneously in the brain. One process tracks the spatial location of the object while the other determines what it is. These functions are so important that there are separate pathways in the brain devoted to each. The "what" information gets parsed through the temporal lobe whereas the "where" information flows over the top of the brain in the parietal lobe. Functional MRI studies in humans show that distant objects in extrapersonal space are processed in the topmost portion of the parietal lobe and near objects in peripersonal space in the lower portion.[4] Along the way, these what and where streams of information communicate with each other, building a neural map of objects in the space around an animal.

Although these brain maps are built using incoming sensory information, they have an intimate connection to the animal's motor system. A cow, or a dog for that matter, does not have labels for anything. It can only relate to things in its world by the nature of its interactions with them. As Hediger realized, everything in an animal's space is something to either acquire or avoid. When primates see objects that are graspable, neurons in the part of the brain that control hand movements begin to fire in coordination with their counterparts in the parietal lobe. So when neurons in a cow's parietal lobe are firing in response to something entering the pasture, motor neurons are also ramping up in preparation for potential action. This could be for approaching if the object is nonthreatening, running away if it is dangerous, licking if it is edible, or bunting if it is movable. In this manner, perception of objects in personal space is heavily influenced by the motor system in "top-down" interactions.

Maybe because they were mommas, Lucy and Ethel were the most obvious monitors of extrapersonal space. Ricky Bobby probably kept an eye out, too, but he seemed to regard fewer things as immediate threats and carried on with his business, which consisted mostly of eating and loafing. Lucy, with calf in tow, had to keep tabs not only on potential threats in her space but also their positions relative to Xena.

Since nothing materializes out of thin air, any intrusion into Lucy's space had to occur by movement into it. This was why she was so attuned to motion. Research has shown that motion is the principal driver of activity in the parts of the brain tasked with monitoring personal space. Two types of motion are likely to trigger the defensive map. Looming stimuli—when something gets bigger in size—indicate movement toward the animal. And speed of movement of an object. The faster the motion of an object in the visual field, the more likely it is to be near and potentially dangerous.

So I made efforts to behave the opposite way: with slow movements and no direct approaches. That was how I gained access to Lucy's extrapersonal space and eventually her peripersonal space. To avoid looming, I learned to make myself small. I approached the cows in a crouch, doing a duck-walk. In this way, my apparent size stayed constant even though I was getting closer. And the closer I got, the more slowly I moved. An optical illusion, really, to give the appearance of unchanging, or at least slowly changing, distance.

In contrast to extrapersonal space, we know a lot more about what constitutes peripersonal space than when Hediger and his peers first defined it. The recent spate of research has taken a dual-pronged approach, focusing on behavioral manipulations that cause PPS to expand and contract as well as neuroscience data demonstrating how PPS is represented in the brain. Almost all of this work has been conducted in humans and monkeys. None has been done in livestock, so we must be cautious in generalizing from species that can grasp things to those that can't. Nevertheless, given the commonalities of brain systems among mammals, we can still use these findings to understand cattle behavior.

Peripersonal space has four key features.[5] First, PPS is anchored to the body and the body parts. When the body moves, so does the PPS. Second, PPS is multisensory. Objects and events within the PPS can be detected by any and all the senses. Third, anything within the PPS is considered actionable. And finally, the PPS is flexible, expanding and contracting as the circumstances dictate.

The biggest debate about peripersonal space is how to reconcile defensive and nondefensive functions. Hediger thought of personal space as a defensive buffer zone against threats, and that is clearly an overriding function. But humans and monkeys also have a nondefensive bubble, within an arm's length radius, surrounding the body. This is the zone in which the animal can directly act upon objects. The nondefensive PPS expands and contracts. Using a tool, for example, extends the reach and expands the PPS. When driving, a human's PPS expands to encompass the car itself. Of course, the car body serves both defensive and nondefensive functions. Think of pushing an object or towing something.

The defensive functions of PPS are well documented in terms of stimuli that cause a startle response. A puff of air near the face, for example, evokes a classic startle response in people and monkeys. The eyes blink and the hands flutter up defensively. Moreover, this response has been localized to two regions in the brain. When the ventral intraparietal area (VIP) and the polysensory zone (PZ) of monkeys were electrically stimulated, the monkeys had the same reflexive movements as they did to air puffs.[6] FMRI studies in humans have shown that analogous brain regions are activated when faces loom in one's PPS.[7] Researchers disagree whether there are separate representations in the brain for defensive and nondefensive PPS. A single representation of objects in one's PPS has the advantage of simplicity. Whether a defensive or nondefensive action is called for is determined by other parts of the motor system. Alternatively, an animal might keep separate representations of threatening and nonthreatening objects in its PPS.

Why does this matter? Cattle are notoriously skittish creatures. An incursion into their PPS, even if it is for something as benign as offering a treat, has the potential for triggering a defensive response. Conversely, when routine veterinary procedures need to be done, like vaccinations, it is advantageous to maintain the animal in a nondefensive state, even though a needle is clearly a threat. The key question is whether a cow keeps separate maps of "good" and "bad" objects in its PPS, or whether there is only one map and the cow has to decide whether to approach or run. The data from primates suggests the former—there are at least two maps in the brain. But cows are not primates. There are more threatening things to a cow than nonthreatening things.

Simplicity suggests only one map in the bovine brain: everything is a threat until proven otherwise.

Temple Grandin called cattle's peripersonal space the *pressure zone* because as a person approaches the PPS, the cow experiences psychological pressure to move away.[8] Stockmen take advantage of this to move cattle, but I needed to teach my herd to let me into their PPS without retreating.

Seeing as the same types of movement—looming and fast motion—that triggered defensive maps in EPS also triggered them in PPS, I continued my sloth impersonations. Ricky Bobby was the easiest to acclimate to my presence. He already enjoyed his neck being scratched, and as long as I approached slowly and came in from below his head, he obliged by sticking out his neck and resting his head on my shoulder while I scratched his brisket. Ethel was the next to come around, perhaps because she saw how Ricky Bobby behaved. She was more sensitive, though. She did not like me hovering over her. When I reached out to scratch her chin, I had to do so even more slowly than I did with Ricky Bobby. For many months, I left Lucy alone. Any motion toward her PPS just caused her to scuttle away.

Our breakthrough came on a late winter evening, six months after the cows had arrived. I had loaded up the hayrack and moved the cows to the big pasture, as was our routine every evening. BB had long stopped trying to get himself stuck in the rack, but I still hung around. The twilight was the most peaceful time of day. A quiet settled over the farm as the daytime critters bedded down for the night, while the nocturnal ones had not yet begun their peregrinations.

With the cows' bellies full of hay, the evening grooming rituals ensued. Ethel would begin by licking Ricky Bobby's neck. With her attending to the king, I did not need to worry about him horning me from behind.

I was about to sidle up to BB when I noticed Lucy waggling her head at me. It was a slow, friendly wag, which the cows used when they wanted one of the others to lick them. I crouched down and approached as slowly as possible. She didn't retreat. Kneeling on the ground, my forehead almost touching her nose, I slowly extended a hand and touched her dewlap. She flinched and took a half step back but didn't run away.

"It's okay," I said in a soft, low voice.

As I continued stroking her neck, the tension in Lucy's body melted away. She bowed her head and we touched forehead to forehead. Her lips parted with a smack and her tongue reached out to give me a tentative lick on the cheek. My heart fluttered like it was a first kiss, which it was, of course. It had taken half a year, but something in Lucy's brain had moved me out of her defensive map and let me penetrate her PPS all the way to the most personal space: her skin.

The skin is, literally, the last line of defense for any animal. It is the boundary between me and not-me. When Lucy finally let me touch her, we embarked on a new phase in our relationship. Unlike Xena and BB, who were accustomed to my touch from the day they were born, my contact was a new experience for Lucy, whom I guessed had never willingly submitted to human touch before.

It is easy to take touch for granted, but if you have ever lost sensation in a part of your body, then you realize just how important touch is. And although we humans are predominately visual animals, we can more readily adapt to the loss of sight than the loss of sensation.

There is no single type of touch. The skin is bedazzled with many different types of tactile sensors, each served by specialized neurons that convert environmental signals into electrical impulses. As in the EPS and PPS, some of these signals are for defense and some are not.

Pain is the most obvious sensation in the defensive category. Step on something sharp and the body withdraws the limb reflexively without any intervention from the brain. Nociceptors are responsible for this type of signal. Although most nociceptors are located in the skin for obvious defensive functions, they are also found in visceral organs, like the digestive tract and bladder, where they signal potential internal dysfunctions. There are three broad types of nociceptors, each responsive to a different type of noxious stimulus. Mechanical nociceptors respond to physical energy and forces that break the skin. Thermal nociceptors react to heat and cold. And chemical nociceptors are activated by a wide range of compounds associated with tissue damage. These can be exogenous, like acid on the skin, or endogenous, from chemicals like histamine that are released when cells are damaged.

Although pain begins with nociception, they are not the same thing. Pain is the brain's interpretation of the signals coming from the nociceptors, and there are many factors that modulate this interpretation. Stress is a big one. When humans are stressed or anxious, noxious stimuli can be perceived as more painful. In cattle, stress appears to release hormones that may dull their response to pain (stress-induced analgesia) because after being stressed, it takes a more noxious stimulus to provoke a reaction.[9] Complicating matters is that cows, like most animals, will mask their pain because distress is a dead giveaway to predators. For this reason, it was long thought that cows did not feel pain. Although veterinarians and cattlemen are beginning to realize this is not true, the FDA has approved only one drug for pain control in cattle and only under specific conditions.

When I touched Lucy, her skin rippled, suggesting it was painful to her.

Actually, it was a reflexive response triggered by a different type of tactile sensor, one responsive to light touch. Cattle, along with horses and even cats and dogs, have a skin reflex for ejecting potential parasites. The panniculus reflex is triggered by light touch, as when a fly lands, but only from the neck down. It is as involuntary as the knee-jerk response. Although Lucy's skin rippling did not, by itself, indicate that my touch hurt her, I found it disconcerting to be repelled like a parasite. So I scratched harder. When the cows groomed each other, they used a lot of force with their very rough tongues. Hard pressure activates a different type of tactile sensor, which doesn't trigger the skin reflex. By applying firm pressure when stroking her neck, I simply duplicated what the cows did.

My relationship with Lucy continued to evolve over the next six months, with her allowing me increased access to her personal space. Some days were better than others. If she wasn't in the mood, Lucy would adopt one of her defensive postures. With the first-level warning, she put her head down and shook it back and forth. This was clearly different than the relaxed waggle that meant *Pet me*. If I didn't heed her warning, then she engaged the second-level defense of bucking. She didn't buck like a rodeo bull; her movement was more like a bunny hop with a little kick of the hind legs, often accompanied by a snort. As a last resort, she would just run away, but rarely did it get to that point. I learned to read her body signals and only made attempts at friendship when she was already relaxed. Having a full belly helped too.

With time, Lucy let me graduate from neck scratches so I could touch her farther away from her head. Around the one-year anniversary of her arrival, she was letting me stroke her hindquarters and even her udder. I made no attempts to milk her, but just getting to second base was a major accomplishment. Having access to Lucy's udder proved useful the following year, when her second calf needed to be weaned.

Had I not spent a year working in Lucy's personal space I would not have gotten close to her udder. And while human contact is part of the daily routine for dairy cows, beef cattle can go their whole lives without a person touching them, except, perhaps, for a vaccination. Lucy showed me that a cow's willingness to let a person into her space was not a matter of genetics. It was a learned behavior. Once she learned I was not a threat, she let me into her circle of trust.

CHAPTER 11

Whistle Training

You don't have to say anything, and you don't have to do anything. Not a thing. Oh, maybe just whistle.

—Lauren Bacall, *To Have and Have Not* (1944)

Gaining access to the cows' personal space was a milestone for sure. And while I worked on gaining Lucy's trust, the others, especially the calves—Xena and BB—would invade *my* personal space without provocation. BB's peripersonal space was especially porous. As soon as he saw me, he would trot over and crane his neck through the fence or rub my leg if I was in the pasture. I was awash with cow attention. But all this free lovin' did not mean the cows would listen to me when I wanted them to do things. They still had a job to do. Granted, it was to do what comes naturally—eating grass—but there were plenty of times when I needed the cows to move from one pasture to another or allow routine veterinary procedures to be done to them.

If I could teach a dog to lie still in an MRI scanner, surely I could teach a cow a few tricks. But how do you train a cow?

This may strike many folks as a strange question, as cows are not considered particularly intelligent or trainable. That's just because most people have not had the opportunity to interact with cows. From a distance, cattle don't appear to be doing much of anything. But even when they're lying down, cattle are constantly monitoring their environment, watching for animals and things that represent potential threats. They have to hold in memory a great deal of information and adapt when changes occur, which is the essence of learning.

Training is just guided learning. To understand how to train an animal, we must understand how they learn, a process that depends on motivation. In order to learn something—that is, make a connection between something in the environment and a requisite action—something must motivate the animal. Dogs, for example, do not generally watch television. Even though they are exposed to TV constantly, they will take notice only if there is another animal on screen, and then only rarely. Dogs have no reason to watch human drama, and they certainly don't understand our humor. But for a treat, they will jump through hoops.

Animal learning boils down to positive and negative incentives—the carrot and the stick. If you can figure out what motivates an animal, then you can teach it to do almost anything within its behavioral repertoire. These principles have been known for over a century, ever since Ivan Pavlov discovered classical conditioning, and they were later elaborated by the behavioral psychologists Edward Thorndike and B. F. Skinner. The techniques work with dogs, and to a lesser degree with cats. They work with horses and pigs. They work with seals and dolphins and elephants. They even work with chickens, with camps devoted to teaching chickens to do all sorts of tricks. (Actually, the camps are designed to teach people how to train animals more generally. The chickens are just the subjects.) And these techniques work with cattle. It just takes longer and more patience than with dogs.

The terminology of behavioral psychology can be confusing, but the ideas are simple to understand. There are two dimensions relevant to training animals. The first dimension describes the nature of the incentive—whether it tends to increase or decrease a behavior. If it increases a behavior, it is called a *reinforcer*. If it decreases a behavior, then it is called a *punishment*. The second dimension of how animals learn describes whether an incentive is given or taken away. This is where the terminology gets confusing. When something is given, it is deemed *positive*; taken away, *negative*. Just like addition and subtraction.

These two dimensions, each with two elements, result in the four canonical types of learning. *Positive reinforcement* occurs when an animal receives something it likes—positive because something has been given that reinforces the behavior that led to the reward. Positive reinforcers include the obvious things related to survival—food and water—but also the company of fellow animals and kindly humans. Similarly, *positive punishment* is the delivery of something unpleasant, which has the effect of decreasing the behavior that preceded the punishment. Punishments are anything that cause physical or emotional distress. *Negative reinforcement* is the removal of something unpleasant. It can be the removal of pain, or it can be as simple as a person leaving a cow's peripersonal space. The removal of whatever psychological pressure the cow experiences is itself rewarding and tends to reinforce the behavior that preceded the removal of pressure. It is a key concept in cattle handling. Finally, *negative punishment* is the removal of something desirable. It is like putting a child in a time-out for bad behavior. Theoretically, it should decrease the occurrence of the bad behavior, but as any parent can attest, negative punishment is probably the least efficient form of learning.

Modern animal training relies heavily on positive reinforcement. The idea is to give animals something they like as a reward for doing what the trainer wants. It keeps the animal in an upbeat frame of mind and motivated to learn. Positive punishments, like whips and cattle prods, are often used in the cattle industry because they are perceived to produce quick results. But they also have the predictable effect of putting animals on edge and are ethically dubious. One must ask, what justification is there for hurting an animal?

So, just as I had done with training dogs for MRI, I set out to use only positive reinforcement to train the cows.

The first step was to figure out what motivated them. Dogs had been easy. A dog never turned down a treat, especially a morsel of sausage, but the cows were surprisingly fickle. They had a limitless supply of grass, so whatever I offered had to be more appealing than the pasture. It turns out that cows have quite the sweet tooth. While some may eat fruits, like bananas and apples, they really go crazy for cookies and bread and cake. Not exactly the healthiest treats. Fortunately, I didn't have to break out the junk food because my herd was happy to gobble down sweet feed, grain coated with molasses. The old man had used it to call the cows, a tradition I had continued from day one.

The problem with using sweet feed for training was that it was hard to mete out in a precise manner. And I couldn't carry it in my pockets because the cows would follow me around licking my jeans. That was cute at first, but after a few frustrated headbutts from Ricky Bobby, I decided that becoming a walking food trough was not in my best interest.

The magic treats the cows seemingly could not get enough of were cattle cubes. The cylinders of pressed alfalfa did not look very appealing to me, but the cows loved them. Ricky Bobby would beg for them by sticking out his tongue and waving it around until I placed a cube on it. Lucy and Ethel bobbed their heads up and down when they wanted some. The cubes also had the advantage of being deliverable in a precise quantity. I could pour a handful in someone's mouth or I could give a single cube. Delivering single cubes would prevent the cows from getting satiated too quickly and losing motivation for whatever I wanted them to do.

The cattle cubes worked great as a reward, but in animal training, timing is everything. Dogs need to be rewarded within three seconds of doing whatever the trainer is teaching them. Any longer and the dog won't make the connection. I assumed that cows had a similar window for reinforcement. This wouldn't be a problem as long as I was standing next to them and had cubes handy. But there is a better way to deliver rewards with precise timing, even from a distance.

Sweet feed and cattle cubes are *primary reinforcers*, things that all animals come hardwired to like. For training purposes, food is the go-to primary reward. The downside, as just noted, is that it has to be delivered right away, which is not always convenient and has the side effect of diverting the animal's attention from what is being taught to where the food is coming from. If you're holding a treat in your hand, it's a sure bet the animal will track your hand with laser focus and might not pay attention to the lesson.

Instead of giving food directly as a reward, most animal trainers use what psychologists call a *conditioned reinforcer* or *secondary reinforcer*. In Pavlov's dog-training experiments, the ringing of a bell was paired with the delivery of food. After time, the dogs began salivating to the bell. The bell was a secondary reinforcer; by itself, it had no reward value to the dogs, but it wields the same motivating power as a primary reinforcer because of the learned association to food. Consider money, which is a conditioned reinforcer for humans. By itself, money can't be used for anything, except perhaps for heat when you burn it. The value of money derives from its ability to be exchanged for other rewards. This process has to be learned—that is, conditioned.

Animal trainers don't use bells anymore. It's all clickers and whistles. Clickers are not very loud and work great when the handler is close to the animal, as is often the case in dog training. But there are many situations where you need to train an animal at a distance, like teaching a border collie to herd sheep. For teaching at a distance, whistles are hard to beat. They are even used to teach tricks to dolphins and sea lions. And because a whistle is held in the mouth, both of the trainer's hands are free to handle the animal and deliver treats when the time comes.

Before you can use a whistle to train an animal, you first have to teach the animal that whenever they hear the whistle, they will get a treat. This is the easy part, no different than what Pavlov did. Although you can use any whistle, ones manufactured specifically for animal training are easily held between the teeth. They also come in different frequencies, which is an advantage when you need to train multiple animals and want to have a different frequency for each.

I grabbed one of the dog whistles left over from the Dog Project and stuck it on a lanyard to hang from my neck when it wasn't between my teeth. It was a little black torpedo about three inches long and made a high-pitched tone (5.7 kilohertz to be precise, or the high F on a piano keyboard). The manufacturer claimed the sound traveled up to a mile—perfect for calling cows from a distance.

Whereas the other cows viewed anything new with suspicion, BB's natural curiosity made him the ideal subject. One morning, when I had the cows up at the barn for some cattle cubes, I caught BB's eye and motioned him to follow me into a stall. I blew the whistle softly. His ears pivoted forward, but he didn't back away. I immediately shoved a cattle cube into his mouth. I blew again and gave him another cube. We repeated this over and over. Ricky Bobby, sensing that treats were being dispensed, bullied his way into the stall too. In short order I was whistling and treating the two boys as fast as I could. The girls—Lucy, Ethel, and Xena—were not so trusting and stayed out of the mix.

We did whistle training morning and evening. Anytime I was dispensing cattle cubes, I was whistling. Pretty soon all the cows were forming a scrum around me when they heard the whistle, so it was safe to assume the association had been made. The whistle had become a conditioned reinforcer. The group training, though, created a conundrum. With all of them jostling for treats, there was no way to direct the whistle at any individual. When I blew the whistle, I tried looking directly at the cow I was intending to reward, but of course everyone else heard the whistle too. Whoever was last in line to get a treat would become impatient. The boys would bunt my thigh in impatience while Ethel smacked her lips. Lucy would gently horn me if she wasn't attended to in time.

With the whistle serving as a secondary reinforcer, I could then use it to teach the cows to do something. If cows were dogs, what would be considered basic obedience? Come when called, for sure. The cows already came running to the whistle, so I could check that off the list. How about leash training? That could be really useful. In livestock handling, the equipment isn't a leash; it's a lead. If I could walk one of the cows on a lead, then maybe the others would follow.

BB was the obvious candidate for lead training, which would be a two-step process. First, I needed to train him to wear something I could attach a lead to. Then I could teach him to walk beside me on a lead.

I suppose I could have used a collar, but because a cow tends to move in the direction its head is pointing, a halter makes more sense. Made from either nylon straps or rope, a cattle halter has two loops. One goes around the muzzle and is connected to another that loops around the back of the head, behind the ears. Fancy halters, like what horses get, are made out of leather. A ring underneath the chin allows for clipping in a lead. With a well-fit halter, you can use the lead to point the cow's head in any direction.

The first day, I just showed BB the halter. He was curious and a bit frightened; his eyes widened and showed the whites. Before he could decide to flee, I blew the whistle and gave him a cattle cube. I could see the wheels turning as his eyes relaxed. *Hmm, maybe not so bad,* he was thinking. I moved it closer. As long as BB didn't back up, I whistled and gave him a treat. And that was the end of the lesson.

We repeated the exercise every day, but each time, I increased the requirement for reward. At first, the reward was for simply not backing away. By about the fifth day, I only rewarded BB if he moved toward the halter and touched it. After that, I started feeding him the cattle cubes through the muzzle loop. When he got used to that, I rewarded him for not withdrawing from the loop. This took another week of daily sessions. Finally, when he was comfortable with his muzzle in the loop, I brought the other strap around the back of his head, loosely at first, and then gradually increased the tension until I could thread it through the buckle. Voilà! BB was halter trained. A cowboy would say he was *halter broke*, but I did it without breaking anybody.

My old dog-training partner was fond of saying, "Slow and steady wins the race," and that is what I did with BB. Sure, it took several weeks, but so what? In BB's mind, he made the choice to put his head in the halter, which is the foundation of positive reinforcement training.

The other cows recognized something odd about BB when he was wearing the halter. Granted, they already pushed him around, but when he trotted around the corral wearing the red halter, Ricky Bobby gave him side-eye while Lucy and Ethel just stayed away. They clearly wanted no part of it. That was okay. I had no way of knowing what experiences the OGs had had before coming to the farm. Maybe someday I would halter train them, too, but it was easier to start with a naive calf like BB.

With BB halter trained, it was then a simple matter to clip a lead to the chin ring. We walked around the barn and the corral. I preferred him to walk on my left side—just like a dog is supposed to do—because I could hold the lead in my left hand while feeding him cattle cubes with my right. I reinforced good walking behavior by whistling and treating as we went. BB loved the attention and the constant stream of treats. He often backed away from the group sessions because Ricky Bobby and Lucy bunted him out of the way, so the halter and lead became a special thing between the two of us. He'd walk right beside me, glued to my left hip, arching his tongue around to accept the treats.

While the whistle training was crucial to achieving a well-behaved herd, I quickly learned that even positive reinforcement had its limitations. True, BB was halter trained, but because he was at the bottom of the social hierarchy, none of the other cows respected him. I had mistakenly assumed that once BB was walking on lead, the others would follow. Apparently not. Ideally, I would have trained Ricky Bobby, too, but every time I showed him the halter, he put his head down and shook it emphatically.

If I was to perform basic veterinary procedures like vaccinations, I would have to find alternative methods to work with the cattle. It was time to put on my cowboy boots and learn some stockmanship skills.

CHAPTER 12

Stockmanship

Raising cattle builds character.
Working them builds vocabulary.

–T-shirt slogan

No cow is going to get away from me.
She doesn't live long enough.

–Bud Williams

BB's halter training was yet another milestone, and I was particularly proud that it had been accomplished with positive reinforcement. But hugs and cattle cubes only went so far. There were many times, like moving the herd between pastures or doing vaccinations, when I had to take charge and become a proper stockman. This was a long-term process, involving as much intuition as science, with many mistakes along the way.

The bermudagrass in the big pasture had been dormant since mid-October, so I had been moving the herd on a biweekly rotation between the three other pastures. The ryegrass in these pastures was six inches tall, which was high enough for grazing, although not thick enough to sustain them. The herd needed supplemental hay and grain for an energy boost. Another problem was that only one of these pastures had a loafing shed, to which the cows needed access when it rained. I couldn't just move the herd to a pasture and leave them without shelter.

I could move the grain trough and hayrack to whichever pasture the cows were in, but this did not solve the shelter problem. Alternatively, I could bring them back to the barn every evening. All through the summer, the cows had become accustomed to getting their grain at the barn, so it seemed natural to continue the routine: put them in the appropriate pasture in the morning for grazing, bring them back to the barn at sunset for grain, and then move them into the big pasture, where the hayrack and loafing shed were located.

As long as I toted the red coffee can filled with cattle cubes, the herd followed me wherever I went. BB was usually the first in line. He trotted on my left, maintaining physical contact, arching his neck around the front of my hip, and sticking out his tongue for a steady stream of cubes. Everyone else just fell in line.

Two of the pastures did not have fence lines connected to the barn area, leaving a gap in fencing when we moved to those locations. The move to the front pasture wasn't too bad as there was only a fifty-foot stretch without fence. It was a straight shot from corral gate to pasture gate, so I strung two ropes, which created a visual alley for the cows. Because of the orientation of their eyes, the cows were drawn into chutes with features parallel to the direction of movement. Perpendicular features, whether cattle grates or lines painted to look like grates, stopped them dead in their tracks and turned them around.

The ropes worked like a charm. With a can full of cattle cubes, the herd followed me through the rope alley and flowed into the pasture.

The baby pasture—so named because it was only one acre—posed a more challenging traverse. It was located five hundred feet from the barn, down a gravel road and across a cement driveway perpendicular to the direction of travel. Cows needed to have sure footing at all times, and the gravel caused them anxiety. At least there was a fence next to the road, leaving a two-foot strip of grass for the cows to walk on. But there was no getting around the cement drive.

The first time I attempted this move, I filled the coffee can and shook it in front of Ricky Bobby. Where his highness went, the others followed. He stuck out his tongue, and I shoved a handful of cubes in his mouth before setting off toward the baby pasture.

The cows exhibited a strong adherence to Newton's first law of motion—that is, an inertial tendency to keep moving in whatever direction they were going. To increase their momentum, I started jogging. So did the herd. I made sure to lead them onto the narrow strip of grass, avoiding the gravel. As we approached the driveway, I picked up speed, figuring with all that momentum they wouldn't screech to a halt. The cement caused Ricky Bobby to change his gait, but he stayed the course and followed me into the pasture. Lucy, bringing up the rear, was the only one who balked at the cement. Her hooves were overgrown, and hard surfaces caused her discomfort. But the pain of being separated from the herd was greater than that of cement under hooves, and she reluctantly crossed over to join her mates. Kathleen followed from behind, ready to provide backup should any of the cows decide to go off course. At night we reversed the process, but the cows needed no additional incentive to return to the barn. They knew well that a trough of grain awaited them and headed home at a gallop.

After a week of this routine, Kathleen had to go to work and wasn't around for backup. The cows knew what was expected of them, so I felt confident I could handle the move on my own.

Big mistake.

Maybe it was a coincidence or, more likely, the cows sensed the absence of a human behind them. When BB got to the gravel drive, he decided to make a sharp left turn and trot off toward the side of the barn facing the main road. I watched helplessly as the entire herd decided that they, too, wanted to investigate what BB had discovered and followed him into unfenced parts of the farm.

I shook the coffee can to no avail. They were not interested.

On the edge of panic, I called Pastor Ken.

He picked up right away, as he always did. "Hey neighbor! How you doin' this fine mornin'?"

A conversation with Ken required a certain time commitment. He made it a point to value every day on this planet and share in his joy of life on a farm. Always the pastor, every conversation began with a sort of homily.

But this was an emergency. I hated being rude, but I had to cut him off before he got rolling. "Ken!" I exclaimed. "The cows have got loose!"

He understood. "I'm on my way."

Although cows thrive on routine, when given the opportunity to explore new pastures, they kick up their heels—literally. Lucy and Ethel, who almost never got excited, were beside themselves jumping up and down, kicking their hind legs in joy. When I moved closer to round them up and turn them back toward the baby pasture, they just moved farther away. I had visions of them wandering onto the main road and getting smacked by a truck, leaving a wreckage of steel and bovine body parts.

While waiting for Ken, I managed to lure everyone except BB into the baby pasture. I'd shake the can and he would come over, but as soon as I opened the gate to put him with the herd, someone else would scoot out.

Ken came barreling across the field and parked his truck across the main driveway, blocking the exit to the road. He climbed down and sized up the mess. BB and Xena were having a time, while their mommas lowed unhappily at them. Unfazed, he said, "Looks like we got some Houdinis."

A second hand made all the difference. I called the rascals and shook the can of cubes, while Ken walked in a sweeping arc behind the calves. The combination of pushing and pulling worked. The herd was reunited, and they galloped off to graze.

The moral of the story was to avoid working cattle alone. Teamwork was the safest, sanest approach, even with miniature cows, but not always practical. I couldn't call Ken every time the cows needed moving. Can one person work cattle? Yes, but it requires a deep understanding of cow psychology—and the right equipment.

In hindsight, three factors contributed to my loss of control over the herd. First, Kathleen's absence removed the psychological pressure on the rear of the cows' peripersonal space. A basic principle of stockmanship has the handler moving toward an animal's PPS, causing the animal to move away.[1] Once the animal is headed in the desired direction, the handler backs off, releasing psychological pressure. The invasion of PPS is perceived as aversive, so the removal of pressure is itself rewarding to the animal, a perfect example of negative reinforcement because the removal of something unpleasant reinforces an action. Over time, the animal learns to anticipate the handler's movements and move in a way that will avoid any pressure at all. Without Kathleen, BB saw that there was no possibility of pressure from behind and was therefore free to move about.

Second, I had assumed that the prospect of cattle cubes was sufficiently rewarding to keep the cows moving toward me. In contrast to the removal of something unpleasant, the provision of a reward for a desired behavior is positive reinforcement. That was how I trained dogs to go in the MRI, and I assumed it would work for teaching cattle. Positive reinforcement is, in fact, a very effective training approach, but it only works if the animal finds what you are offering rewarding. I had yet to meet a dog that didn't find treats rewarding, but cows were not so consistent. Bud Williams, a legend among stockmen, famously said, "Calling animals and using feed to take 'em works great except that sometimes they're not hungry, and then it just don't work."[2] A handful of cattle cubes couldn't compete with wide-open pastures.

Third, by relying entirely on positive and negative reinforcement, I had failed to provide any independent visual guidance of where the cows should go. There was a fence line on one side of the intended trajectory, but the other direction was open. I should have used ropes to draw the cows' attention along the path, like I had with the move to the front pasture. This was an easy fix and worked wonders on subsequent traverses.

These three factors—pushing, pulling, and visual guidance—formed the foundation of stockmanship at the farm. Experienced cattlemen would find none of this surprising, but to a neuroscientist, all of it was endlessly fascinating. Working cattle wasn't just a matter of pragmatics. I wanted to know *why* the cows did what they did. When I understood the whyfors, then I could become a better stockman and work the herd in a way that was more harmonious for all of us.

When it comes to working cattle, it's not long before the name Bud Williams comes up. Williams was born on a farm in Oregon in 1932. When he was twenty, he and his wife, Eunice, began working cattle and sheep on ranches throughout Northern California. Williams quickly acquired a reputation for being the cowboy who could deal with difficult cattle. He learned to rotationally graze cattle on the open western ranges without using fences, just by teaching the cattle to stay together as a herd where he wanted them to go. In 1989 Williams began teaching his unique methods to other people. The heart of his approach was to let the cattle teach you what they want and how to use that knowledge. Eunice once said of Bud, "His great powers are observation and pure stubbornness."[3] Williams died of pancreatic cancer in 2012.

In the ideal handling situation, the stockman positions himself and uses minimal movements to encourage the herd to make choices that align with what the stockman wants. In other words, get the cattle to think it's their choice of where to go. Forcing them only causes stress and takes ten times longer than if they are relaxed. Williams's mantra was "Slow is fast."

If all I had to do was move the herd from one pasture to another, I could get away with minimal handling. A bucket of cattle cubes and Kathleen as backup would be all that was necessary. But, of course, that wasn't all there was to being a cattle rancher. There were vaccinations to be done. During fly season, insecticide needed to be applied to the cows' backs. Sometimes they got sick and needed to be restrained for the vet. And then there was the annual hoof trimming. As the herd continued to grow, it seemed like something always needed to be done with one of them.

The cows had a preternatural sense for when something was up. Any stranger, especially one who had the stink of other animals' fear on them, set them on edge. No amount of cattle cubes, sweet feed, or sweet talk could get them out of the state of anxiety the prospect of procedures put them into. When Lucy, Ethel, and Ricky Bobby first came to the farm, the vet came out to vaccinate them. With no cattle handling facilities, the best I could do was capture them in a horse stall. They crammed themselves into a corner, trembling in fear as the vet approached with her syringes. She was blessedly quick and jabbed each of them in the rump as they tried to scoot by. Ricky Bobby was the worst, trying to leap over the stall gate. It's a terrifying sight to see a bull with his front legs over a fence in naked panic.

In hindsight, I had made many mistakes the sages of stockmanship could have helped me avoid. I did not yet possess that wisdom. But with each mistake, I made course corrections and began to improve my stockmanship. It was clear I needed to provide the cows with facilities in which they were comfortable and could be handled safely for routine care.

First up was some type of facility in which they could be restrained. Cows are not like horses. You can't just put a halter on a cow and expect it to stand there while you doctor it. For that matter, horses won't do that either, unless they have been trained.

Cattle benefit from a chute. The industry standard is an alleyway

twenty-six inches wide, which is just big enough for an adult cow or steer to squeeze through without allowing it to turn around. Thanks to Temple Grandin's insights, the ideal chute (or *race* as they call it in the UK) curves in a gentle arc so the cows can't see what's at the end.[4] They just follow the animal in front of them.

Implementing a full-on Temple Grandin design was not in the farm budget. My first attempt was to add a fence line next to the barn stall. I set it twenty-six inches from the wall and figured that it would serve as a poor man's cattle chute. If only I could have gotten the cows to go into it. BB was the only one who could be lured into the chute, and even then, it was iffy.

I resolved to do better. This was when I learned of Bud Williams and his cattle-handling techniques. He is the only person to have a type of cattle-handling facility named after him. The *Bud box* is deceptively simple. Even to experienced cattle-men unfamiliar with the concept, it looks like a standard rectangular holding pen with gates at each end. An alley is positioned in the middle of one side, heading off at right angles. An interior gate can be swung flush with the alley. The cow walks into the Bud box and heads to the back gate. Once it sees there is no way out, it heads back to where it came from.

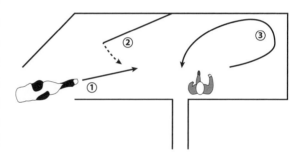

Bud box. (1) The cow enters the pen and heads toward the back wall. (2) As the cow passes the alley, the handler closes the interior gate. (3) The cow turns around and heads back to where it came from but, with the interior gate closed, flows around the handler into the alley.

The handler, though, swings closed the middle gate and positions himself next to the alley. The cow then flows around him into the chute.

The description sounds crazy, but I needed to try something, so I ordered a bunch of cattle panels to build a Bud box. If it didn't work, I could always repurpose the panels into some other configuration.

I placed the Bud box in the corral, next to the gate leading to the big pasture. The cows were used to coming and going through this gate twice a day, so they would have to walk through the Bud box. The first few times I left the gates open at both ends, just to get them used to the box. On the third day, I did as Williams described and closed the far gate. Just as predicted, they walked to the far end and, finding the exit blocked, turned around. I pulled shut the interior gate and stood next to the alleyway. The cows flowed around me and out. No big deal to them.

It was a big deal to me, though. I had struggled for months to contain the cows safely for vaccines and such. In just a few days, with a change in equipment and positioning myself in the correct manner, the cows walked into the alley with zero drama. Once they were in the alley, I could close both ends and the cows would stand there quietly. No roping, no running around chasing cows. I could even do this without Kathleen as backup.

Despite the Bud box being so effective, it is not in common use across the livestock industry. The apparatus looks like a holding pen, and that's how many cattlemen would treat it. Williams, though, was adamant that cows should never be held in it. The box is just a pass-through to turn the cows around and go in the alleyway. The handler must position himself correctly, applying and releasing psychological pressure at just the right moments. In other words, the Bud box does not work on its own. It requires a handler who understands cow psychology.

Williams's approach to cattle handling depended on three principles.[5] First, the stockman should employ techniques that cattle respond to naturally. This means you should not rely on techniques that require animals to learn something new. Cows can learn many things, and over time, learned behaviors may become part of the stockman's repertoire, like my herd learning to come to the shake of a coffee can. The problem with learned behaviors, as opposed to natural behaviors, is that cows are individuals, and not all will learn equally. As I discovered with BB, it only takes one stray cow to pull the herd in a different direction.

What is a natural behavior? Perhaps the most important to a stockman is the cow's tendency to move in the direction its head is pointing. If you can control that, then you've done half your job. If you enter their peripersonal space, or pressure zone, and the cow doesn't want you there, its natural tendency is to move away in the direction it is already pointing. The stockman can turn a cow by getting her to look in a different direction. Because a cow doesn't like people directly behind her in the blind spot, if you stand to one side, behind a hip, the cow will turn her head slightly to get a better look. When a stockman applies psychological pressure to the hip, the cow will naturally change her trajectory and curve to that side.

Second, the stockman must not use force. Cattle should be allowed to do what you want them to do. Force only leads to fear and stress. According to the first principle, frightened cattle will revert to naturally protective behaviors. They will seek the protection of the herd and attempt to return to places of known safety. If they can't, panic ensues, and they may stampede or attempt to break through fencing. The solution is to constrain the set of choices cattle have. This is the fundamental insight into the use of the Bud box. As the cows move into the box, the handler removes the choice of leaving through the entrance. The only remaining option is to leave through the alley. Done correctly, the cows think it's their choice.

Finally, the stockman should avoid doing things that bother cattle. This means no yelling. No cattle prods. No flags or sticks. No pressure from directly behind. And no fast movements. This last point reiterates Williams's mantra: "Slow is fast."

As a novice cattleman, I was naturally eager to get procedures over with quickly. The early vaccination episodes had reinforced the unpleasantness for me and the cows. *Best to be done with them as fast as possible*, I thought. Over time, and after learning Williams's methods, I came to appreciate the wisdom in slowing down. The neuroscience research had shown that fast movements into the peripersonal space triggered defensive movements. When a cow is stressed—either by something painful or by separation from the herd—the hormone cortisol is released from the adrenal glands. Cortisol affects almost every system of the body, readying it for fight or flight—mostly the latter for cattle. Heart rate increases. Bowels are expressed. Muscles tense. Studies in cattle have shown that cortisol levels peak ten to twenty minutes after a stressor and don't return to baseline for forty minutes.[6] Once triggered, it could take up to an hour for a cow to reset from the defensive mode. It seems obvious that it is far better, and more efficient, to move slowly and avoid triggering the cow's cortisol release.

It wasn't until the third season that I began to gain some mastery of the zen of taking one's time.

Ricky Bobby and Lucy needed their hooves trimmed. Cattlemen aren't usually concerned about hoof trimming because most beef cows don't live long enough for their hooves to get overgrown. Dairy cows are another story. But there weren't many dairy farms in middle Georgia, so finding someone who knew how to care for cattle hooves was a challenge. Horse farriers wouldn't touch a cow. Apart from the fact that most cows would not stand still for the job, cattle have two hooves on each foot where horses have only one. The biomechanics are completely different.

Even with the Bud box, this was not a job that could be done easily in the alleyway. Restraining the cow's head would not prevent her from kicking. Cattle had to be completely restrained, immobilizing the legs—both for the animal's safety and the trimmer's. There was only one person in the entire state who had the equipment and knowledge to trim cattle hooves.

Chad Boyce grew up on a dairy farm, where he learned how to care for cows' hooves. After taking over the family business, he realized that hoof trimming was a rare skill and he could make a better living doing that than running a dairy. So he sold the cows and assembled a portable restraint system that he towed behind his truck as he traveled around the Southeast, tending to the hooves of dairy cows and the occasional family cow.

On a crisp spring morning, Chad pulled up and eyed the cows, who were milling about the corral.

"I haven't had a lot of success with the minis," he said. "Did one where the vet had to completely sedate it."

"I've improved my handling system," I said hopefully.

Chad looked at the Bud box skeptically. "You gonna get them in there?"

"I can," I replied. "Do you want to set up first?"

He shook his head. "There's all sorts of smells on the equipment. Blood, hair, nails. Once I pull in, they won't go near it."

I took a deep breath to settle myself and get into the cow zone. As casually as possible, I ambled into the corral. The cows, of course, knew something was up. They were starting to get a little agitated, bunching into a corner. They hadn't yet reached the panic stage and so were still amenable to following my direction.

I approached the group slowly from the side. Just as Williams described, they started to move away in the direction they were pointing. Luckily, they were already aimed toward the Bud box. As I moved closer, they scooted by, and because they were used to going through the box, they headed right in. I followed and casually pulled the gate of the box closed behind me. With that exit removed, the group flowed into the alley. First BB then Lucy, Ricky Bobby, and Ethel. I closed the slider behind them and let them chill out for a bit while Chad set up. With the alley full, I let Xena run free in the pasture.

The key to doing the job was Chad's tilt table. This was a cattle chute that could hold one cow. Two large straps, attached to a hydraulic lift, went under the belly and lifted the cow off the ground. Once airborne, the entire unit tilted horizontally. With the cow on its side and legs sticking out, Chad could get to work with a grinding wheel on the hooves.

BB took some coaxing to walk through the contraption, but once in place we discovered he wasn't quite long enough to straddle the straps, so we let him go. Lucy took all of this in stride. She stuck her head through the gate at the end and didn't even struggle when Chad lifted her off the ground and tilted the table. I wondered if she understood this would make her feet feel better. Five minutes of work on each hoof and she was done. She trotted off like a ballet dancer *en pointe*.

And that left Ricky Bobby. Poor Ricky Bobby. True, he was the bull. But I knew him to be a weenie to the core. He had backed up into the chute as far as possible from the instrument of torture. He stood there trembling slightly, shit all over himself and the bars. Chad was trying to nudge him from behind to move him forward.

Slow is fast.

"Let's give him some time to settle down," I said.

Chad backed off.

I crouched down and began stroking Ricky Bobby's dewlap. "It's okay, big boy," I cooed softly.

After about five minutes, the tension in his body had noticeably decreased. I stood up and walked slowly toward the lift table. Ricky Bobby followed. When he got to the end, he reared up and tried to squeeze through the opening of the headgate, but all that accomplished was putting his head right where we wanted it. With his head captured, he couldn't move forward or backward. Chad strapped him in and lifted the man-baby up and onto his side. I think Ricky Bobby realized that in this position, resistance was futile. He relaxed, and Chad gave him a cow mani-pedi.

The Bud box had been eye-opening. It made possible all the routine care that needed to be done, without spending a fortune on an elaborate cattle-handling facility. It worked because I understood something about cattle psychology and because the cows had come to trust me. I doubt that it would have been as successful if I had tried Williams's approach when I first got the cows. His method of handling was as philosophical as practical. The stockman had to understand what the cattle wanted, which was not apparent to a novice. The cows, in turn, had to learn to trust their stockman. Williams had an aura that made cattle comfortable. They would follow him anywhere after only an hour of getting to know each other.

Stockmanship is an important aspect of raising livestock that focuses on the relationship between cattle and their human handlers. But what do cows do when there are no humans around? Do they have their own set of rules—a cow code of conduct? Without a human to tell them where to go, how do they decide where to graze? Where to sleep? As a neuroscientist, I was curious.

The more time I spent with the cows, the more convinced I became that, left to their own devices, cattle society operated according to a complex set of rules and relationships between herd members. A human's presence distorted the natural structure, making it hard to discern what the cows did for their own sake versus what they did for the stockman.

How could I observe the cows without letting them know they were being watched?

CHAPTER 13

Cow Cam

A camera is a tool for learning how
to see without a camera.

–Dorothea Lange

As much as I would have liked to spend more time
with the cows, keeping up with the other farmwork
demanded my attention. Fences needed mending. The garden needing tending. And the pastures still needed mowing.
At the end of the day, I just wanted to rest. Wouldn't it be
nice to keep tabs on the cows without having to go out to the
pasture? Then I could see what they did when I wasn't there. But
spying on the cows wasn't the only reason for a cow cam. With
advances in image recognition, it had become possible to program
a system to recognize the cows and analyze what they were doing.
Behavioral analysis, group dynamics, even improving pasture
management were all possible.

Putting up the cameras was the easy part. That was simply a
matter of purchasing equipment that had a high resolution and
long focal length to reach the far corners of the bigger pastures.
The barn served as the camera hub. I put up some masts built
from PVC pipe and mounted cameras pointing in different
directions to get the maximum coverage. A wireless bridge

to the main house let me stream the feed from the barn. I repurposed an old computer to run the software and store recordings.

The feed was endlessly fascinating. I checked it first thing in the morning to see if the cows were up yet. Usually they weren't. They didn't start their daily routine until I had had my second cup of coffee. Maybe this was because they were so well fed that they could take their time to start grazing, or maybe they had learned my schedule and didn't bother coming up to the barn until they knew I would be there. I also checked the cow cam before I went to bed. On nights with a full moon, I could make out the ghostly shapes of the cows when they moved. When they were close to the cameras, the infrared light would reflect off the tapetum in their eyes. Multiple pairs of glowing dots would appear and wink out as they turned their heads.

As near as I could tell, the cows never slept. Obviously this couldn't be true, as all animals need to sleep. But if they were sleeping, they were doing it on the sly. There was not much research on cattle's sleep habits. One study of dairy cows found that cows spent thirteen hours a day awake and eight hours ruminating.[1] The remaining three hours were divided between various stages of sleep, from drowsing to REM and non-REM sleep. However, even the paltry three hours of actual sleep weren't continuous. Bouts of sleep lasted only four to five minutes. My herd's sleep habits appeared consistent with this, and although they showed no signs of sleep deprivation, it made me a bit sad to think about this aspect of cows' lives. They were hardwired to be on alert, snatching bits of sleep here and there.

The cow cam really proved its worth in the second season, when Lucy and Ethel calved again. Cattle gestation, on average, is 283 days—identical to the average human pregnancy of forty weeks. A cow's estrous cycle is also similar to that of humans, although the cow's is faster, coming every twenty-one days. In order to become pregnant, the bull must do his business in the twenty-four hours just prior to ovulation. During this period, the cow becomes receptive to the bull's entreaties. She becomes restless and may mount the other cows. When she's ready, she will stand in front of the bull with her tail out, which is why it's called *standing heat*. It is not subtle. They may go at it over and over again for several hours. The cycle repeats every three weeks until she is pregnant. If four weeks pass without another standing heat, then you know the cow is pregnant. Both Lucy and Ethel had

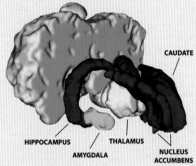

▲ Cutaway rendering of parts of the limbic system of the cow.

◄ Lucy and Princess Xena. Lucy stays close to Xena, who is two weeks old. (Julie Nestor)

▼ The two calves, Princess Xena and B.B., became fast friends. Ethel stands behind them while Ricky Bobby relaxes. (Gregory Berns)

▲ Lucy licking B.B. He is adopting the stereotypical pose of outstretched neck, indicating he likes it. This was not a common occurrence, as they were unrelated. (Gregory Berns)

▼ B.B. asks for a belly rub by rolling on his side. (Gregory Berns)

▲ Lucy's calf nursing.
Note the similarity to B.B.'s
posture during neck licking.
(Gregory Berns)

◀ B.B. reciprocates
by licking me.
(Gregory Berns)

◀ Synchronized grazing. All the cows are oriented in the same direction and moving together as they graze. Ricky Bobby is still lying down and hasn't yet joined the line. (Gregory Berns)

◀ Relaxed grazing. The herd's collective momentum has decreased, and individual cows are grazing in random directions. (Gregory Berns)

◀ Cow cam output. One video frame with the output of the neural net identifying the locations of the cows, along with the confidence level that each is, in fact, a cow. (Gregory Berns)

▲ Human view (left) versus cow's view (right) of an apple slice being offered to Ethel. (Gregory Berns)

▼ Ricky Bobby getting his hooves trimmed. Although he is not happy with the procedure, he is relatively relaxed. The alleyway from the Bud box is in the background. (Gregory Berns)

▲ Lucy's newborn calf, here just three minutes old. Xena stands behind Lucy, watching what she will have to do herself. Ricky Bobby has just scented the new calf and is exhibiting a flehmen reaction by curling his lip. Tex is about to realize that his world is upended, while Daisy takes it all in stride. (Gregory Berns)

▼ Lucy grooming Cricket while everyone minds their own business—except Tex, who stares at them from a safe distance. (Gregory Berns)

▲ The entire herd has fun with a pile of sand. (Gregory Berns)

◀ Cricket, two months old, engaging in locomotor play by herself. She is simultaneously running, jumping, kicking up her heels, and twisting her head. (Gregory Berns)

◀ Cow version of a play bow. Walker (left) is signaling he wants to play with Cricket (right) by putting his head down, holding his butt high in the air, and wagging his tail. Although his head is down, his neck is straight in line with his back, which indicates playful intent (as opposed to aggression, where the chin is tucked into the chest). (Gregory Berns)

▲ The herd was fascinated by their reflections. Tex and B.B. gaze intently.
(Gregory Berns)

▼ Cow session with Jerome and Ken. (Gregory Berns)

been bred on their second heat cycle after Xena and BB had been born, so as their due dates drew closer, I was ready.

Instead of doing cow-checks for signs of labor every couple of hours, I could keep tabs on what was happening through the cow cam. It was early June, and the heat index was climbing past one hundred every afternoon. Lucy had been springing for a few days, so the calf was coming soon. Three days before her due date, I checked the stream after morning coffee and saw the herd had formed a scrum in the middle of the pasture. This was an unusual formation, so I headed out to see what was going on. Sure enough, Lucy was standing proudly over a beautiful chestnut calf. She had already licked it clean. No doubt about the sex this time. The little guy had a healthy pair of nuggets.

I called Ken with the happy news, and he came right over to see the new member of the herd.

For the next hour we watched as Lucy first got the calf on his feet and then started pushing him back to her udder. The little guy had a strong rooting reflex. He tried to latch on to Lucy's legs, but with her encouragement he eventually found his way to her udder, which looked as if it would start gushing milk as soon as he touched it. When he figured it out, he really went to town, draining one side in no time.

BB sensed that milk was flowing and tried to sneak in, too, but Lucy firmly shooed him away.

Ken laughed. "BB's saying, 'Can I work on one of those, too, while we're at it?'"

I named the new calf Texas Ranger—Tex for short—because as any fan of *Talladega Nights* knows, Ricky Bobby had two sons: Walker and Texas Ranger.

Three weeks later, Ethel calved. This time, I was eating lunch while monitoring the cow cam. Ethel had separated herself from the herd and was pacing around the pasture when I caught a glimpse of something poking out of her rear. By the time I got on my boots and made it out there, the calf was already on the ground, covered in yellow goo. Once again, Kathleen, Ken, and I went through the routine of staying with them until we were sure the calf was latched on and nursing. This time it was a little heifer. We named her Daisy Duke. At least the calvings were getting easier.

Although the cow cam was useful for checking on what the cows were up to, watching real-time streams still made it difficult to see the big picture of patterns of behavior. In this sense, the stream was no different than standing in the pasture and observing the cows. For the most part, they moved at a glacial pace, which was a timescale too slow for human brains to appreciate. The real power of the cow cam was revealed in the recordings, which could be speeded up. Instead of watching the cows for two hours at a stretch, I could fast-forward the recording in two minutes. Here, the cows appeared to move at a human pace, almost jogging through the pastures. And seeing the cows moving at human speed meant the human mind had an easier time discerning patterns of movement.

It was obvious that the cows did not move randomly through the pasture. They followed established trajectories, tracing patterns in the pasture like a spirograph. And not just patterns in space—patterns in time. Their morning grazing routine was different than in the afternoon. Their travels were consistent and followed a diurnal rhythm. The cows were locked into their own strange attractor.

Through the cow cam recordings, I was able to identify two broad grazing patterns. Most of the time, the cows engaged in what could be called *synchronized grazing*. In this pattern, the herd members stayed close to one another and, crucially, oriented themselves in the same direction. A common configuration would have the herd lined up shoulder to shoulder as they grazed. Because cows move in the direction their heads are facing, the herd would move en masse through the pasture, like the offensive line of a football team. This was grazing with a purpose.

The other pattern was *relaxed grazing*. This tended to occur after a period of synchronized grazing, when the cows began to get full. The line of movement would typically slow down, and some of the members might pair off and engage in some mutual grooming. With the collective momentum decreased, individual cows would shift their orientation and graze this way and that but still orbit around the herd's center of gravity, which often coincided with Ricky Bobby's location. Bud Williams described this as a settled herd.[2]

The existence of these two grazing modes was consistent with recent advances in the computer analysis of collective animal behavior. Iain Couzin, a professor at the Max Planck Institute of Animal Behavior in Konstanz, Germany, had pioneered this type of analysis over the last several decades.[3]

Couzin and his colleagues had developed software for tracking individual animals in large groups and analyzing their motions to understand how group behavior emerged. For example, a flock of birds might appear to have a mind of its own. But, of course, there was no central mind controlling the flock. It only appeared that way through the coordinated movements of all the individuals. How, then, did the individuals know what to do? That was the question that had puzzled animal scientists for centuries and to which Couzin had devoted his career to answering.

Couzin suggested that animals that engage in collective behavior have two competing motivations. The first is an individual's ability and motivation to explore the environment in pursuit of what they desire. For cattle, this manifests as a willingness to strike off on one's own to find the best forage. It aligned closely with the personality dimension of openness, for which BB and Ricky Bobby scored the highest. The exploratory motivation was counteracted by the second motivation, what Couzin called *sociality*. This is an animal's tendency to align its direction of travel with nearby individuals. There is a great evolutionary advantage to this strategy. The law of large numbers tells us that any individual in a group is likely to be wrong about where the best forage lies. For this reason, an exploratory animal takes a risk by striking out on his own. The group consensus—the average of each individual's opinion—is very likely to converge on the best forage. Therefore, individuals are almost always better off doing what everyone else is doing. They don't even need to look at the whole herd. If an animal does what its neighbors are doing, and they are doing what their neighbors are doing, the herd will end up aligning in the same direction without even knowing it. This was Couzin's great insight: that collective behavior emerges through simple rules of individual behavior.

The cow cam demonstrated that the herd followed these principles exactly as Couzin's model predicted. When the cows engaged in synchronized grazing, they did so without looking up to check what the others were doing. All they needed to do was monitor the direction of travel of their nearest neighbors, which they could do perfectly fine with head down, while grazing grass.

These two motivations—exploration and sociality—need not be constant. As cows fill up their rumens, their motivation to forage decreases and the evolutionary imperative to do what the others are doing also decreases.

From a cow's perspective, if you're already full, there isn't anything to lose by striking off on your own to see if there is something tastier to eat. The collective motion disperses and individual cows graze in random directions.

Couzin's early models treated a herd as a collection of individuals that were essentially identical to one another. This was a simplifying assumption to make the computer models easier to work with. As the field advanced, Couzin and his colleagues incorporated more realistic scenarios in which individuals were not all the same, what was called *heterogeneity*. I could see plenty of heterogeneity in my own herd, but when it came to herd dynamics, it was not what one might have expected.

The obvious assumption would be that Ricky Bobby—the bull—was the leader of the herd. This was true in some ways but not others. When Ricky Bobby wanted something, whether it was access to the hayrack, a bucket of grain, or a choice spot to lie down in, he bunted everyone else out of the way. And when he really got cranking, he was the proverbial bull in the china shop, flipping anything and anybody that got in his way. But the cow cam showed that he very rarely led the herd in synchronized grazing. Quite the opposite. He was often the last one to join in. Rather, it was usually one of the adult cows—Lucy or Ethel—who led the graze. Later work by Couzin and his colleagues offered a simple explanation.

An individual with a high nutritional requirement will often assume a leading position because she has a higher motivation to find the best forage.[4] This is a riskier strategy than following one's neighbors, but the reward is potentially finding the best food and eating it first. A lactating cow has the highest energy needs of a herd—more than a bull of the same weight. A pregnant cow that hasn't started nursing doesn't require as much food, but she still needs more than, say, a steer or a heifer. This would explain why Lucy and Ethel were frequently the leaders of synchronized grazing. Ricky Bobby, at least when it came to food, turned out to be rather risk averse and stuck to sure things like making a beeline to cattle cubes or following the ladies when grazing.

One of the predictions of Couzin's model was that herds would spontaneously split apart and reform. The fission-fusion process occurs when differences of opinion arise. When two potential leaders begin to move in different directions, the herd will tend to graze in the average direction of motion. When the difference of opinion becomes too great, a herd will often split, with some following one leader and some following the other. It reforms when either the leaders align their opinions or when one of the subgroups breaks up and gravitates back to the other subgroup. However, I never observed my herd splitting. Although they might disperse during relaxed grazing, never did they split into two groups. The most likely reason was that the herd was not big enough to split. The calves didn't count because they had to go wherever their mommas went.

While these computer models did a good job of explaining the grazing behavior of the herd, they weren't designed to simulate how a herd decided to lie down and ruminate. The same general principles should apply, except instead of individuals deciding which direction to forage, they would need to decide where to rest. It was, perhaps, similar to Couzin's example of how a colony of ants decided where to build a nest. When one cow found a good place to rest, the other cows would see that and take note of how comfortable the individual appeared. Then, as others joined him, the process accelerated through positive feedback. The bigger the group got, the more it attracted the others.

Even though he rarely led the herd in synchronized grazing, Ricky Bobby was usually the first to lie down. For this reason, and because of his size and patriarchal stature in the herd, he acted as an anchoring point for when the others decided that they, too, were done grazing and ready to ruminate. Ricky Bobby seemed particularly popular with the young'uns. Tex and Daisy would often settle in next to him. I found this oddly touching that the calves preferred to cozy up to their papa over their mommas.

In the cows' choice of where to settle down to rest or ruminate, different factors came into play than when the cows were grazing. The clumping of the cows during rest was driven by both attractive and repulsive forces. As a prey species, the safety afforded by the group acted to draw members together. Their vulnerability was greatest when resting because even though a cow could run away, that extra delay caused by getting to her feet could make all the difference in the world. This was why the cows clumped more tightly together when they rested than when they grazed. But this attraction was not without bounds. When too many cows bunched together, it became uncomfortable. The cow cam showed that they had fitful rest. The cows slept in snatches of only five or ten minutes. They would get up and reposition themselves, and, as BB and Ricky Bobby were fond of doing, they sometimes flopped on their sides and stretched out their legs. Anyone lying too close to them would get a sharp hoof in the side. These sorts of movements tended to space them apart.

The cow cam footage provided a wealth of information about the herd dynamics, but it was tedious to scroll through the recordings. And even when I found something of interest, it was still merely a snapshot in time. To really have practical utility, the cow cam needed a backend: an AI that could sift through the frames and identify the cows. With that information, I could see the precise patterns in space and time that the cows traced out.

Automated image recognition had become so commonplace that even the cameras I had installed came with software that could identify common objects in real time. They did not, however, allow for an easy way to store that information for later analysis. It was better to add image recognition on the backend, after the video was recorded. There were several versions of this type of software that were freely available. All relied on neural nets.

A neural net is a computer algorithm that simulates, in a fashion, how the brain processes information. The magic is in how the neural net is constructed, or, as computer scientists say, its architecture. For image recognition, a neural net takes as its input a 2D image, with each pixel representing a simulated neuron. Each of these neurons is connected to the neurons of another layer, where the incoming information is processed and sent on to the next layer, and so on. A typical architecture for image recognition has hundreds of layers of neurons, arranged like a massive chocolate torte. The final layer—the output layer—might contain anywhere from two to one hundred neurons, each representing a discrete type of object, like person, dog, car, or cow. To make all of this work, the net has to be trained, which requires thousands, often hundreds of thousands, of images. For each training image, the net is told what the image is, and then the strengths of the neuron-neuron connections are tweaked slightly to move the output toward that answer. The bigger the training set, the better the net becomes. Training these types of neural nets takes a lot of time and computer resources. Fortunately, there are several pretrained nets that, with a little computer programming, can be deployed in any application one can imagine.

Mostly it was a matter of getting all the ancillary programs and libraries installed that allowed the neural net to work. Because these nets are computationally intensive, they run faster using the resources of the computer's graphics card. Getting that to work required more fiddling. After several months of trial and error and scouring the web for help, I finally had a working AI backend. It wasn't as fast I would have liked, taking about a second to process one video frame, but the AI found the cows and recorded their locations as they moved about the pastures.[5]

To really see what was happening, I converted the cows' locations in the video frames to GPS coordinates and projected them onto a satellite image. It's surprising what you can see with a bird's-eye view. I could now see the actual trajectories through the pasture without the foreshortening effect of watching from the fence line (which also occurred with the cameras). Couzin's theory was remarkably accurate. The herd movements looked similar to those of a flock of birds. And my observation that the herd never split was proved wrong. Viewed from above, it was apparent that the herd did, occasionally, split into two subgroups. These fissures were usually short lived, and the subgroups inevitably merged, but they showed that the cows sometimes had differences in opinion of the best direction to graze, and the herd voted with their feet.

The tracks also showed the subtle effects of the fence line. Cows often move single file along fence lines, leaving well-trodden paths. This is not desirable for efficient pasture utilization. The grass gets stamped into oblivion, and the bare dirt becomes prone to erosion. There could be many reasons the cows do this. It could be force of habit because fence lines usually lead to food or water. It could be that as prey animals, cows have an innate fear of the open and prefer to hug the boundary for safety. Rodents do this, tracing out paths along baseboards. Or it could be that fences act as visual guides, drawing the cows along the direction of the rails, just like what happens when they are sucked into the alleyway of the Bud box. I suspected this might be the case because the GPS tracks showed the cows following trajectories that paralleled the curvature of the fence line even when they weren't right up against it.

These sorts of analyses reveal a lot about the hidden lives of cows. Much of what cattle do is invisible to us because of their glacial pace. Casual observations suggest they are relatively inert animals that don't do much of anything, but, of course, this isn't true. The cow cam recordings, sped up and fed through a neural network, uncovered a wealth of complex interactions between the cows and their environment. This new type of knowledge could have profound implications for how cattle are raised.

The cow cam could be used to track the interactions within the herd. Probably the most important interaction to a farmer is determining which cows have been bred. You can't watch a bull 24/7, but a cow cam can. You can even train the AI to recognize breeding events. In the new era of farming, AI won't mean only *artificial insemination*. It will be *artificial intelligence*. It remains to be seen whether this new AI will be used exclusively to increase production efficiency or will also be used to improve the welfare of the animals.

From a production standpoint, farmers want to use their pastures in the most efficient way possible. Management-intensive grazing (MIG) is exactly that—demanding the farmer subdivide his land into small paddocks and move the cattle on a daily basis to ensure complete consumption of the forage, not returning for several weeks. It's a lot of work, and many farmers would rather stick with letting cattle graze large pastures and supplementing with hay in the winter, even if it means wasted forage. But MIG need not be so labor intensive. Bud Williams was famous for getting herds to graze where he wanted without the need for fences or hotwire. For him, it was a matter of leading cattle to the desired location and getting them settled, or what I called *relaxed grazing*.

The cow cam revealed how much time the cows were spending in each part of the pastures. If they were spending too much time in one area, I put a protein tub where I wanted them to spend more time. The tub was like a cow-sized lollipop of molasses and other nutrients and exerted a pull on the cows to return to it throughout the day.

The cow cam had revealed many aspects of the cows' behavior that weren't apparent even from intense observation. The practical applications of the video stream ranged from monitoring cow labor to quantifying herd movement and grazing patterns. But the cow cam also made clear how dynamic the individuals were. The cows were hardly passive participants in the farm environment. They were active, sentient beings in their own right.

CHAPTER 14

Ricky Bobby Holds a Grudge

Forget injuries. Never forget kindness.

—Confucius

They say elephants never forget, but neither do cows.

By the summer of the second year, after Tex and Daisy were born, Ricky Bobby's horns had nearly doubled in length. I had assumed he had been fully grown when I bought him, but in retrospect, he had probably been only two years old. In the intervening year, his horns had gone from a manageable six inches to foot-long instruments of torment. This had begun to present problems on multiple fronts. Like all the cows, Ricky Bobby used his head to express his mood. If someone or something was in his way, he bunted them aside. Where Lucy's and Ethel's horns grew forward and upward in graceful arcs—what some call devil horns—Ricky Bobby's grew outward from the sides of his head, like a miniature Texas longhorn.

When his wingspan exceeded the width of his head, Ricky Bobby began to wield the increase in his personal space like a mercurial dictator. If he was in an impatient mood, a flick of his head cleared any nearby cows from his path. If I crouched down to give someone else neck scratches, Ricky Bobby would saunter over, wagging his head to move the others aside.

Ricky Bobby soon discovered that he could move large objects with his horns. After the evening grain was consumed, he would hook a horn under the grain trough and flip it over in frustration. I also had to stake the hayrack to the ground because the man-baby kept turning it over until it wedged against the fence.

All of this bull work took a toll on his horns. Where they were once smooth pillars of keratin, they became gnarled trunks. Shards flaked off and little spears pointed this way and that, like a medieval mace. Some of these hangnails went all the way to the base of the horn, where the leverage on them began to peel back the skin. Not only was this a hazard to me and the other cows, but it was also a potential source of infection. Unlike deer antlers, which are shed every season, cattle horns are permanent and become contiguous with the skull. The stray bits of horn needed to be clipped.

A pair of goat hoof trimmers would get the job done. After all, horns and hoofs are made from the same material. My target was a spear about six inches long that pointed out at almost a right angle from the base of the horn. One day, while Ricky Bobby was lying down, chewing his cud, I approached him as I would if I were going to lie down with him. He gave me side-eye as I slowly pulled out the trimmers. I inched toward the offending spear, targeting its base to snip it clean. With the clippers in position, I squeezed. And . . . nothing. The hangnail was much tougher than I had expected.

Ricky Bobby realized what was happening and wanted nothing to do with it. It's surprising how fast a bull can get to his feet. He was up before I could disengage the clippers, the sudden movement yanking on the hangnail, which only irritated it more. I'd have to try again later.

The next evening, I approached him. He eyed me suspiciously, so I held out both hands. "See? Nothing in my hands."

He wasn't buying it. Ricky Bobby trained one eye on my right pocket, where the clippers had been hidden. He knew they were there. Fool me once, right? But not twice. He turned away and walked off.

This went on for several days. Finally, I gave up. He would have to work off the flakes of horn by himself. With all the head butting and flipping of objects, the spears eventually broke off at the base. Sometimes they bled for a bit. I chalked it up to standard bull behavior.

What I found interesting was how good Ricky Bobby's memory was. He remembered that I had hidden the clippers in my pocket. You'd think it would be a case of out of sight, out of mind. But not for him. Ricky Bobby had demonstrated object permanence, just as Piaget's infants had.

Although there was an overall relationship between brain size and performance on the A-not-B test, subsequent attempts to measure object permanence in ungulates had not yielded consistent answers. The performance of goats was variable, with only some figuring out the A-not-B test.[1] Alpacas didn't fare much better, although some could learn to go to the B-location with enough practice.[2] In a study of giraffes, bison, and buffalos, the giraffes and bison performed above chance, but the buffalos didn't.[3]

It was surprising that so many exotic species had been subjected to these tests of cognitive function yet cattle hadn't. It was just another instance of scientists exhibiting the same types of biases that most people had against cows, assuming that they were dumb, uninteresting animals. Out of all the animals that had been tested on the A-not-B test, bison were the most closely related to cows. If bison could squeak by on the test, it was reasonable to assume that cows could too.

I suspect that most animals have object permanence, and the failures of some species on the A-not-B test had more to do with the challenges of teaching an animal what they were expected to do with the apparatus than with the absence of object permanence. It is often difficult to convey to an animal how they are to register a response on these tests. So it is with cows. Ricky Bobby's behavior showed that he had kept a mental representation of the hoof trimmers. Even when they were hidden in my pocket, he knew they were there.

Lucy and Ethel had shown evidence of object permanence, too, although it happened in very different circumstances. When Xena and BB were still young, they would chase each other around the corral in the evening. While their mommas kept a watchful eye, the two calves would weave between the adults, cutting left and right and sometimes dashing into the barn stall. As soon as they disappeared behind the stall wall, their mommas would take notice. They would stop grazing until their calves emerged again. Sometimes they would go over to the wall and look for them. The first time they did this, I assumed the mommas didn't understand that their calves were still there. But by about the third time I witnessed this behavior, it became apparent that Lucy and Ethel knew that Xena and BB were in the stall, because they were pacing back and forth in front of the wall. This was the bovine equivalent of an infant reaching for the toy under the box.

In their frustration, the mommas didn't understand that they needed to go around the wall to get to their calves. The cows did the same thing when they got stuck on the wrong side of the fence. Even though everyone was in plain sight, they didn't seem to understand how to circle back to a gate to join the herd. They just paced back and forth along the fence line, lowing in distress. These navigational errors stemmed from a different cognitive limitation. Although cows have object permanence, they often don't understand how to get around artificial barriers. This is surprising because cattle have excellent spatial navigation and even perform well on simple mazes.[4] Although cows do well on maze tests, they still have to learn to navigate them. Because I hadn't taught the cows specifically how to reach a goal behind a wall or fence, they couldn't figure it out when they were in a tizzy trying to get to their calves.

Cows aren't the only animals that have difficulty with barriers. It is so common that researchers have been studying the phenomenon for over a century using a test called the *detour paradigm*.[5] With either a transparent or opaque barrier, an animal has to figure out how to go around it to reach its goal, usually a food reward. It is the kind of skill humans take for granted, but it does not come naturally for many other animals. Chimpanzees can figure it out without any training, and so can dogs, but most other animals require experience with the apparatus to learn how to navigate around the barrier. Many factors affect an animal's performance on the detour paradigm. Distance, for example: the farther away the goal, the easier it is for an animal to navigate around the barrier. Conversely, neophobic animals have more difficulty than exploratory animals because the neophobic ones are afraid of anything new. The most relevant factor, though, is the reward value of the goal. Counterintuitively, the more powerful the reward, the harder it is for an animal to detour around a barrier. A high-value reward so captures the animal's attention, it can't disengage to look for indirect routes. This seemed the most likely explanation for the cows' behavior when they got stuck on the wrong side of the fence or a calf disappeared behind a stall. They became so fixated on reaching their calves that they couldn't think of looking for alternative paths.

Ricky Bobby, though, had no trouble navigating away from my pocket.

He gave me the cold shoulder for a week. Sure, he took the cattle cubes I offered, but as soon my hands were empty, he would amble off, seeking licks from Ethel, who was always happy to console the man-baby when he was in a funk.

One of the key elements of object permanence is memory. The cows had already demonstrated their innate ability to remember *where* things were in their environment, but Ricky Bobby's grudge against the trimmers showed that he also had a great memory for *what* things were.

Psychologists have long known that there are different types of memory. Spatial memory and object memory, for example. These types of memory are dependent on the hippocampus in the brain. Cows' large hippocampi are necessary for navigating and remembering the contours of their grazing environment, but they also support the memory functions necessary for remembering who is who in the herd. Cows can easily discriminate pictures of herd mates from other cows not in their herd.[6] This is a remarkable skill that even dogs don't demonstrate.

Ricky Bobby's behavior also suggested a different type of memory: a remembrance of the event itself. So-called *episodic memory* is like a snippet of a movie. For humans, episodic memories are the threads that bind together the fabric of a person's life. We use these remembrances to stitch together a coherent narrative of what has happened to us and who we are. Episodic memories give meaning to our lives.

Do other animals have episodic memories? It is a difficult question to answer. Episodic memories require not only thinking about oneself but thinking about oneself in a different time frame. Many researchers believe that only humans can accomplish this mental time travel.[7] A growing body of research, however, has begun to suggest that many animals have the rudiments of mental time travel. Birds and squirrels, for example, remember where and when they cached food for later use as well as what type of food it was.[8] Perhaps more important than what/where/when something happened is the ability to replay the event itself. Rats, for example, can remember a specific sequence of odors, and the replay of the sequence can be disrupted by inactivating the hippocampus.[9] Not everyone agrees that these results indicate that rats can mentally time travel, but I find them compelling enough to suspect that that was what Ricky Bobby was doing when I approached him after the horn-trimming incident.[10]

If cows have episodic memories, then that would mean they also have a sort of narrative in their heads. Obviously these wouldn't be verbal narratives like the inner monologues that people have. But what if cows have a sort of pictorial story of their lives? A visual highlight reel, if you will. I have wondered whether dogs have something similar, remembering snippets of events that were important to them. Humans, dogs, and cows all have the same basic brain structures to support episodic memory. From my perspective as a neuroscientist, it seemed more plausible than the alternative of human exceptionalism that posits that only people are capable of episodic memory.

Even if cows have only fragmentary memories for events, it would be enough to form a base of knowledge about how things worked on the farm. This knowledge—when I fed them; which people were friendly to them; how to behave with one another—could only be acquired through experience. Taken together, each of the cows' experiences, which were stored as episodic memories in their brains, formed a collective knowledge base—what could be called *cow culture*.

The idea of cows having a culture would strike most people as wackadoodle. I didn't bring it up at the seed-and-feed, but Ken and I talked about it frequently. He didn't think cow culture was crazy. Quite the opposite. We agreed that the cows were skilled at picking up what was expected of them and, through the collective mentality of the herd, spread that knowledge to one another, probably by observing one another. These observations of what was acceptable and unacceptable behavior could only be stored as episodic memories.

It wasn't until the third season, though, when the last batch of calves was born, that I saw hints of cow culture. By then, Ricky Bobby had long forgiven me for the horn-trimming episode. He had held his grudge for only a week before returning to his usual needy behaviors of demanding neck scratches.

CHAPTER 15

Too Many Cows

Cows are like potato chips. You can't have just one.

−Riff on a T-shirt/coffee mug slogan

(usually referring to dogs or cats)

Cattle ranchers dislike open cows, those not successfully bred by the bull. I must have had an exceptionally fertile herd because in the third season every female that could have been pregnant was with calf. No open cows here. Lucy and Ethel were carrying their third calves, and Xena was pregnant thanks to my misreading of her standing heat the previous summer. It wasn't the cows' fault. The problem was Ricky Bobby.

Granted, I hadn't fully planned for controlling the cows' reproduction. I had thought that whenever a cow was in heat, I would just remove Ricky Bobby to a separate pasture for a day or two. But when I tried that when Xena went into heat, Ricky Bobby turned into a snorting, grunting, charging bull. He spent day and night trying to find a way through the fence. He didn't eat or drink and ended up with urinary stones as a result. It was all for nothing anyway. After two days of this, I figured Xena was out of heat and let the big boy back with the herd. Wrong. An hour later, Xena stood under the big oak tree with her tail in the air—a sure sign she had just been impregnated. She had been *bred back*, which was a type of artificial selection to concentrate

135

desirable traits by breeding within a line. But I knew the truth. I had messed up, and I needed to get my herd under control.

I had three choices. I could split the herd, separating Ricky Bobby with BB to keep each other company. But seeing how he acted when Xena was in heat, I foresaw broken fences and potential injuries with that strategy. I could sell some cows, which is what a real cattleman would have done and would have covered some of the expenditures the herd was incurring. But who was I kidding? Nobody would care for them like I did, and I couldn't bear the thought of one of them ending up at auction. That left only one option: castrate Ricky Bobby.

Bull calves are routinely castrated, usually during their first week of life. It is just part of the deal in raising cattle. Conventional wisdom says that the sooner it's done, the less traumatic it is for the calf. Banding is one method. A small rubber band is placed around the neck of the scrotum, choking off the blood supply, and the whole thing falls off a few weeks later. Or you can slice open the scrotum and strip out the testicles. The initial trauma is worse than with banding, but the calves heal faster. We had used the surgical approach for BB and Tex. Experienced cattlemen do it themselves, but we had opted to let our local veterinarian handle that procedure.

Castrating an adult bull is not so simple. The blood supply is fully developed, so there is a real risk of hemorrhage. Infection is a threat because the bulls are constantly lying on the ground. The nerve supply is also fully developed, so the pain is worse than what a week-old calf experiences. And there is the bulliness. Getting a naive calf restrained in the chute was child's play compared to coaxing in Ricky Bobby. He would be on high alert as soon as the vet showed up, and the sight of any surgical instrument would send him into a full-blown panic. Although he had forgiven me for the horn trimming experience, he still remembered that surgical instruments caused pain, which was dangerous for both him and the vet. He would need to be sedated.

The date of the operation kept getting pushed back. The vet knew this was not going to be easy and had to set aside a morning for the procedure. We wanted to do it sometime in the winter, when the temperature was cool and no flies would be around to lay eggs in the open wound. But every time we set a date, it rained, or it was too muddy from having just rained. After three postponements, everything aligned on a crisp March morning.

As planned, Kathleen and I put Ricky Bobby in the alley of the Bud box by himself. This wasn't hard because he usually went first. Kathleen closed the head gate and I shut the slider at the other end before the other cows went in. Ricky Bobby didn't like it, but he could hang out there safely until the vet arrived.

It takes a special kind of person to be a large animal veterinarian. The pay is poor, even for a profession in which low salaries are the norm. Farmers are notoriously thrifty and quick to complain about costs. A large animal vet might not have the overhead of a fancy pet clinic—the truck is their clinic—but they don't get compensated for all the time they spend driving around to farms in sparsely populated rural areas. Furthermore, the work is dangerous. Large animal vets have war stories of injuries they've sustained, and most don't deal with cattle, preferring to stick with horses, which are better behaved and more lucrative.

For over thirty years, Melissa Fulton had been running a large animal practice out of a building on her farm on the other side of the county. She served mostly equine patients, but she also tended to cattle. She and her long-time vet tech, Mary Kate Jobe, pulled up in a beat-up pickup outfitted with storage boxes holding all the veterinary accoutrements. Their expressions telegraphed that they would rather be doing something else. They had seen Ricky Bobby on previous visits and knew that although friendly with me, he was terrified of them. The last time Melissa and Mary Kate were there, Ricky Bobby had reared up on his hind legs and powered right through the head gate.

That was not going to happen again. This time he got a little something to calm him down. Melissa called it *K-stun*, a mixture of ketamine and xylazine given in a small enough dose to cause sedation but not so much that the animal lies down. (Technically, it's called *standing K-stun*. A larger dose that causes the animal to lie down is *recumbent K-stun*.)

Twenty minutes later, the big guy was zoned out, saliva dripping from his mouth. We put a halter on him, and Melissa squeezed into the alley to get to work. Mary Kate held his tail over his back, which numbed the area below, while Melissa gave him a local anesthetic. She made quick work of it, although Ricky Bobby did decide to lie down halfway through. A tug on the halter got him back on his feet, and Melissa finished the job. There was surprisingly little blood.

"Just remember," she said, "there's still sperm in the pipeline for two weeks."

The cows weren't due to calve for two months, so there was no possibility of other pregnancies in that time.

"How about testosterone?" I asked.

"That will decrease after a couple of months."

We let him back in the pasture, and all the other cows came running over, forming a huddle around him. Ethel, who seemed to have the strongest bond with him, licked him all over, trying to comfort him. Poor Ricky Bobby. Apart from the expected discomfort during the healing process, which would take about a month, I don't think he was ever really conscious of what had happened. He would always be the bull in his mind—and mine.

With the Ricky Bobby situation dealt with, it was time to turn my attention to the next round of calves. Lucy was first up.

The ideal time to breed a cow is three months after she has given birth. Because the gestation period is nine months, this will result in a calf every twelve months. But Lucy had gone into heat a mere four weeks after giving birth to Tex, and Ricky Bobby had wasted no time. Wham, bam, pregnant on the first go-round. Not ideal for the mother's health, but at least it would move up her calving from the heat of the summer to the milder spring, when the pastures would be lush.

If only for Tex.

Texas Ranger was prepossessed of startlingly good looks: full-bodied, chestnut fur, and a white splotch on his forehead, like his daddy. Oh, and the swagger, like he knew he was destined to be the homecoming king. As Pastor Ken liked to say, "Tex, you're a mess!"

Tex enjoyed testing the patience of the other cows, either by head butting or, more boldly, by mounting them. At his age, this was purely a dominance thing. But when he pushed BB or Ricky Bobby too far and they ran him off, little Tex would seek Lucy's protection and stare back at the others from under her chin, knowing full well nobody would challenge Lucy. Yes, he was quite the momma's boy, and he had been skating through life on his good looks.

I couldn't figure out why Lucy put up with it. She was not a meek cow. If she didn't like something, she made her disapproval clear by snorting and

stamping her feet. She was not above using her horns either. However, she indulged Tex for the spoiled bull calf that he was. This included letting him nurse whenever he wanted, long past the age when he should have been weaned. By the time Tex was nine months old, he was almost as tall as Lucy, and yet in the evening, after I had given them grain, he would bash her udder, forcing the milk to let down. It was like ice cream for dessert.

Calves are normally weaned between six and eight months. Tex was a month past that, and because Lucy had become pregnant so quickly, she was due to give birth again when Tex was ten months old. A yearling and a newborn can't share milk; the older calf will take it all for himself. Lactation takes more energy than being pregnant, and here Lucy was doing both. She was not as fat as she had been when pregnant with Tex. It was obvious that Tex had been siphoning off nutrition from his soon-to-be sibling even before it was born. Even worse, there would be no colostrum. Lucy needed to dry up before she could bag up again.

The simplest solution would have been to separate Tex into a different pasture. I could have put BB with him to keep him company. The herd, though, had become tightly knit. The calves had not known any other way of life, and they took great comfort in the structure the herd provided. Could they have adapted to being separated? Yes, but the price would have been days, or weeks, of bawling. The distress cows feel when separated from their herd, especially from their mommas, is as intense as that of a human child being left at daycare for the first time. Besides, my pasture situation could not accommodate splitting the herd. There weren't enough pastures with good forage to use two simultaneously.

Cattlemen sometimes use a nose flap to wean a calf. This is a piece of plastic that clips into the nose. When the calf tries to nurse, the flap pushes the teat out of the way. The reviews are mixed on how well this works. The flap is meant to be a temporary bridge until the calf is physically separated from the momma. If the calf remains with the momma, it will usually start nursing again as soon as the flap is removed.

I took a different approach. After two years on the farm, Lucy fully trusted me. She let me touch her anywhere, including her udder. To discourage Tex from nursing, I mixed up a dilute tincture of black pepper oil—not concentrated enough to sting but enough to have an unappealing smell. I slathered the mixture over Lucy's udder. Tex didn't like that and sulked away, confused about what had happened to the snack bar. I did this twice a day to make sure the scent stayed fresh. Lucy seemed to understand because she finally started to kick Tex away whenever he began nosing around her hind legs. After a week, the process was complete. Lucy's teats shriveled up and her udder began to shrink.

Tex would occasionally test the waters, but Lucy wouldn't stand for it anymore. She knew the next calf was coming soon.

Four days before Lucy's due date, I was moving the herd to the front pasture. I put up the ropes and led the cows with the coffee can of cattle cubes, as we had done hundreds of times. As usual, Lucy was the last in line, but this time, she stopped halfway and refused to go any farther. She knew her calf was coming and did not want to have it in the front pasture. Can't say I blame her. With five hundred feet of road frontage, a wobbly newborn could slip through the fence rails into the street. What was astonishing was that Lucy knew this too. I took her cue and moved the herd to a different pasture, away from the road.

The next morning, Lucy paced around the pasture, uninterested in eating. By noon she was in obvious discomfort, alternately standing and lying down. Sure enough, around one o'clock that afternoon, a tiny white hoof poked out her rear. Calves (like horses) are born with slippers over their hooves. The *eponychium* is a soft white covering that protects momma from getting injured during the birth. As soon as the calf is born, the exposure to air hardens up the covering, and within minutes, the hoof turns to its pigmented form. But the calf wasn't quite ready to emerge yet. The calf slippers would appear for a few minutes and then retreat.

The labor lasted two hours. At 3:00 p.m. the lower legs were out in the open. Next, the button of a nose appeared beside them, and then the entire muzzle. Lucy's back hunched up as she bore down for one final push. The head popped out and an eye blinked. The forelimbs reached for the ground. Gravity took care of the rest. The body of the calf slid out, doing a half-flip as it landed on its back with a splat.

Lucy attended to it immediately. She used her raspy tongue to clear away the goo from the calf's nose and mouth. I couldn't tell if the calf was breathing, but her legs were twitching, so I knew she was alive. Before I could get a closer look, Lucy had flipped the calf right side up and was licking and horning her with clear intention: *Get up!* The calf, who I could now see was a little heifer, responded. She didn't have her legs under her yet, but she was alert and breathing and no longer splayed out on her back.

The herd came running over, Tex in the lead. Everyone got a sniff of the new calf. Ricky Bobby had a classic flehmen reaction: he curled his upper lip in a wide grin as he sucked the calf's smell through the scent organ in the roof of his mouth, making an olfactory imprint of the new herd member. Poor Tex just stared at his new sister as it dawned on him that he was no longer momma's favorite. His head visibly sagged.

Lucy had the calf on her feet in thirty minutes. The newborn wobbled unsteadily, her elbow joints bending backward. They would stiffen up in a few days. Lucy kept nosing the calf back toward her udder. As soon as the calf's nose made contact with Lucy's fur, it started rooting around for a teat. This is an involuntary reflex all mammals are born with, including humans, but because the calf was on her feet it looked like she knew what to do. She rooted along Lucy's hind leg until she found the teat. A sniff, and then a tentative suck. Just as I had seen with Xena and Tex, the calf's contact with the udder triggered a surge of oxytocin, and Lucy began licking the calf's rear even more furiously, which had the effect of mashing her face into that teat.

The calf worked the teat for fifteen minutes and then moved on to the front one, draining the entire left side of Lucy's udder. Her belly visibly full, the calf backed away. She blinked in the afternoon sunlight and took in her new, wonderful world.

Ken, alerted to the birth, had ambled up silently. "It never gets old," he sighed. Then to the calf, "Oh, you are so lucky to be here on this farm, with such a good momma and a fine herdsman to take care of you."

The calf was healthy but weighed a mere twenty-five pounds. Calves born to normal-sized cows typically weigh seventy-five pounds. There was nothing abnormal about our new calf's weight. It was about the same as Daisy's and Tex's birth weights. But the day after she was born, the temperature dropped from the eighties to the fifties. A breeze made it feel even colder, and the calf started shivering. She did not have enough body mass to thermoregulate.

Ken and I discussed the options. We could put Lucy and the calf in a stall up at the barn, which had fresh straw laid out for just this scenario. It wasn't heated, but it would keep the wind out.

"I don't know," I said. "They don't like being confined, and they're doing so well with the herd."

"That's true," Ken replied. "Can you spread some hay down at the loafing shed to give the calf something to nestle in?"

The loafing shed was located in the lower part of the big pasture in an area that had been carved out of the surrounding forest. It was just two walls and a roof, but the walls faced north and west—the direction of the prevailing winds. The shed provided shelter from the rain and, crucially, the chilling effects of the wind.

I humped three bales of hay down there and spread it to a depth of eight inches. I didn't even have to call the cows. They all came trotting over to see what was going on. You'd think they'd never seen hay before because they started munching down their bedding. Lucy came over with the calf, who was shivering something terrible. I picked her up and placed her in the most sheltered corner. Her legs felt cold as her circulation was being diverted to her core. I buried her up to her neck with hay. Lucy snorted anxiously. I could not have done this a year earlier. Lucy would not have trusted me. But she did this time, and it was the right thing to do.

After ten minutes the shivering lessened, becoming sporadic instead of constant. It was dark, and Ricky Bobby had plopped down in the hay. The rest of the herd would do so soon, so I gave Lucy a neck scratch and told her to take good care of the calf. I texted Ken, asking him to say a prayer for them. And then I headed up to the house for the night.

I gave the calf a fifty-fifty chance of surviving. She faced many challenges. The coyotes were ever present, and the calf would have made a nice meal for them. Every night the first week, the temperature dropped into the low fifties. It rained. The wind blew. But somehow, the heifer not only survived but thrived, putting on ten pounds that first week. At that point, I stopped calling her Fifty-Fifty and named her Cricket—after Cricket O'Dell, a minor character in the *Archie* series of cartoons whose special talent was sniffing out treasure.

The appearance of a new calf had a demonstrable effect on the herd dynamics. Ricky Bobby was still the titular leader, although Lucy was the one who really called the shots. With Lucy busy with her new calf, the others were free to act out in ways that would have previously drawn a stern headshake from Lucy. She made it clear that nobody would get near Cricket without her permission.

It was mostly the yearlings that tested the boundaries. BB became bolder, partly due to his age. He was an adolescent and, like all adolescents, liked to push the others' buttons. Mostly he would goad another cow into a head-to-head pushing contest. If he was feeling particularly lucky, he would even take on Ricky Bobby. Xena, herself pregnant, stayed out of the fray and concentrated on eating as much as she could. She was still achingly affectionate with me, lying down next to me and putting her head in my lap. Daisy Duke was an independent lady who seemed relatively unaffected by the new calf. I think this was because Ethel had kicked her off nursing when she was only four months old.

Tex took Cricket's birth the hardest. Lucy made it abundantly clear that he was no longer welcome at the milk bar. Not only that, he had to keep a safe distance from the new calf. If he got too close, Lucy would run him off. And then Tex would have to stand there, from afar, watching Lucy groom Cricket. The other cows paid no attention. Why should they? But Tex did. If cows could pout, Tex would have had a very long, sad face.

Many times I asked myself if I was just anthropomorphizing these emotions. Ken didn't think so, but he wasn't an impartial observer either. The cow cam showed I wasn't wrong. A sequence of images captured Lucy grooming Cricket. Everyone else was minding their own business—except Tex. Standing beneath the big oak tree, he just stared at Lucy and the calf.

After a week, Tex stopped his sulking, but I can't believe he ever forgot the hurt of being kicked to the curb by his momma. Maybe, someday, when Cricket was older, Lucy would once again give Tex the licks he so clearly missed.

I knew what the skeptics would be thinking. How did I know that Tex experienced the bovine equivalent of jealousy? Perhaps he was watching Lucy and Cricket for some other reason. Maybe he just wanted the milk, which might rise to the level of envy but not jealousy. However, his baleful looks didn't occur when Cricket was nursing. And after the first day, Tex stopped sniffing around Lucy's udder. So I didn't think his behavior was motivated by a desire to regain access to the milk. He was long past the age when he needed the nutrition anyway. I thought his late nursing was more for comfort than anything else, as was the licking he had received from Lucy. The cow cam supported this theory because Tex tended to watch Lucy when she was grooming Cricket, not nursing. None of the other cows paid them any attention, so it was something about the Lucy-Tex-Cricket relationship that made Tex behave the way he did.

Ethel calved three weeks later, just a day past her due date. Ethel's signs of labor were more subtle than Lucy's. Ethel's udder had been slowly bagging up for a week, and she started to get a bit of pudding butt, but her behavior wasn't obviously different. It was raining the day she calved, so I monitored her via cow cam and checked on her every two hours. At noon, the herd was huddled under shelter in the loafing shed. Everyone looked glum, but apparently the discomfort of getting wet was worse than their hunger. So they stood there waiting for the rain to stop. Although I didn't know it at the time, the cow cam recording showed that Ethel had been making brief forays into the pasture. In retrospect, her restlessness was the only sign of labor.

At two o'clock, Kathleen and I went to the barn. The herd came stampeding over, carrying on in a most unusual manner. Several of them were mooing. They rarely vocalized anymore and certainly never more than one at a time. Even Ricky Bobby and BB joined in, rasping out hoarse moos, like old men coughing up their morning phlegm. They were clearly communicating something important to us.

Ethel wasn't there.

We headed down to the loafing shed. Sure enough, Ethel was standing over a white calf. She had licked off most of the birth goo, but the calf was shivering and covered in dirt. The placenta lay nearby in the sand. The herd came running over and formed a scrum around the new herd member. Ethel just grunted plaintively as she tried to get her baby on its feet. I could see it was a little bull, but he was either too weak, too cold, or too stunned to get up and start nursing. If the placenta was out, he must have been born right after my noontime check. At two hours old, he should have been up and nursing.

I fetched a warm bucket of water and toweled off the dirt from the calf. I got him on his feet, and, as he stood there wobbling back and forth, I dipped the umbilical cord in a cup of iodine solution. This helped sterilize it and promote the process of drying up and falling off. Ethel licked it off immediately. My rubbing seemed to stimulate the calf as he started to root around for a teat. Confirming my initial impression that he wasn't the brightest candle in the bunch, he directed his rooting at a ridge in the siding of the loafing shed. Ethel lowed softly in frustration.

Sometimes the babies needed a little help. I tried to redirect the calf in the direction of Ethel's udder. Much to my surprise, Lucy stuck her head into the mix and gently nudged the calf toward her momma. This helped hem in the calf until he figured out what to do. After about an hour, he finally made contact with a teat and took some tentative sucks. But they were too feeble to get the milk flowing. He would have to suck harder. And so he moved on to other teat-like protuberances, like Ethel's ankle.

It was hard to watch, but one thing I had learned over the previous two years was to give the cows time and to have faith in their ability to raise a calf. Ethel and Lucy were seasoned cows. They knew what to do. Ethel kept licking and nosing her calf back to her udder. Lucy stayed nearby in case Ethel needed help. By four o'clock the little bull calf had latched on and drained the two back quarters of Ethel's udder. He had even figured out head bunting, bashing the udder to release more milk.

Seeing as this might be the last bull calf on the farm, we had to name him Walker, to go along with Texas Ranger.

Walker wobbled around for the first three days, as the calves always did. Ethel was a helicopter mom and complained incessantly whenever her calf wandered more than a few feet from her. No wonder Walker quickly learned to ignore his momma's vocalizations, which was evident by his fondness for squeezing under the rails and trotting outside the fence line. This caused Ethel intense distress. All she could do was follow him along, lowing repeatedly until I heard the ruckus and put the calf back in the pasture. This wasn't so simple; he was already as big as Cricket, and when I picked him up he would thrash wildly, bleating like a goat.

Cricket was beside herself with joy. Once Walker got his legs, the two of them became fast friends. During the day, Cricket and Walker would run circles around each other or practice head butting before crashing down for a nap. They would curl up with each other while the adults grazed. Pretty soon the two calves were spending more time with each other than with their mommas, who were always there for milk and protection but not fun and games.

Xena was the last one due. As a heifer, this would be Xena's first calf. I crossed my fingers that the maternal instincts would kick in and that she had learned enough through observation of Lucy and Ethel. By Memorial Day she was tight as a tick. Her udder, although beginning to bag up, was not nearly as big as Lucy's or Ethel's had been the week before they calved. I hoped that Xena would deliver a small heifer without need of assistance.

The day after her official due date, Xena showed the classic signs of a cow in labor. She passed on the morning grain and spent the day following the herd around the pasture but not really grazing. She just picked at the grass.

By late in the afternoon, Xena was well into the first stage of labor. This was when the calf entered the birth canal and the cervix dilated. Cows will often isolate themselves from the herd and repeatedly lie down and get up. Xena was too tightly bonded to the herd to isolate, but she did adopt postures I hadn't seen before. She did the cow version of downward dog—kneeling on her elbows, butt in the air and tail held to the side. Then she would flop down on her side and wriggle like a dog taking a dirt bath. *Gosh*, I thought, *she must be uncomfortable.* Normally, this stage lasted two to six hours, but in a heifer, it could be eight to twelve. Xena probably had started around noon.

By six in the evening, Xena's discomfort was even more intense. No position offered relief, and she continually got up and down and rolled on her side. Of course, as soon as I took a break from observation to eat dinner, Xena went into the second stage of labor. The cow cam captured everything. Two hooves appeared from her rear. Unlike Lucy, who pushed Cricket out from the standing position, Xena squirted out her calf lying down. As soon as the calf was out, Ricky Bobby came running over and began helping Xena clean it off. By the time I got there, just seven minutes after the big event, Xena had the front half of the calf cleaned off. I was grateful her maternal instincts had kicked in and she knew what to do.

The next challenge was getting the calf on its feet and nursing. This

proved more challenging than with any of the other calves. The calf, who I could see was a heifer, was trembling even though it was ninety degrees. *Must be the shock of birth and surge of adrenaline*, I thought. After half an hour, I intervened and stood the calf up. She wobbled precariously but didn't fall over. A good sign. Unlike Cricket, though, this one didn't have a strong rooting reflex. Even when I nudged her toward Xena's udder, the calf just stood there, kind of stuporous. Xena did her best by continuing to lick the calf, but now the entire herd tried to get involved. Too many cooks were in the kitchen.

Kathleen and I were having flashbacks to when Xena was born and the difficulty we had had in getting her to nurse. But since we never saw Xena born, we never really knew whether she had nursed before we found her that night. In hindsight, and six calves later, Lucy had probably taken care of everything. But Xena had never had a calf of her own and lacked Lucy's experience.

After ten minutes of tottering around, the calf's instincts kicked in and she started rooting around for a teat. Not as vigorously as Cricket or Walker had, but good enough. The problem was that Xena kept backing away as soon as the calf made contact with her udder. It looked painfully full, bulging around the teats. An experienced cow knows that she must get the calf to nurse even though it hurts. Xena, even though she had seen Lucy and Ethel do this many times, probably didn't understand why she hurt everywhere. If Lucy hadn't had Cricket to care for, she might have helped out Xena. It was hard to tell with all the jostling going on whether the other cows were trying to hem in Xena and the calf, or whether they were just investigating out of their own curiosity.

Ken showed up just as the sun was setting. He agreed that it was time to help out Xena before it got too dark to see what was happening. So, as I had done with Xena herself two years before, I picked up the calf and moved her to the barn. Xena became frantic and kept circling the area where the calf was born. I had to hold up the calf so she could see it; then she came running, and I was able to confine the two of them.

Ken and I were finally able to get Xena settled by boxing her into a corner of the pen. I fed her a steady stream of cattle cubes while Ken directed the calf to her udder. It took half an hour, but just as it was getting dark, we were rewarded with the sound of a calf suckling. She drained one side of Xena's udder, and her belly bulged full of the critical colostrum. The calf went to sleep.

I opened up the gate to the pasture, figuring that Xena and the calf would stay put for the night. By the time I got back to the house and checked the cow cam, they were gone. Thankfully, there was a full moon, otherwise I would never have found the calf, who was nestled in the tall grass in the middle of the pasture. The herd was in the loafing shed. Poor Xena was torn. She wanted to be with the herd, but she also felt the overpowering urge to tend to her calf. So she ended up pacing back and forth between the herd and calf, bellowing all the while. She would nudge and moo at the calf, wanting her to get up and follow Xena to the herd. As much as I would have liked to help, I dared not move the calf in the dark. It would only confuse Xena and worsen her distress.

The moonlight was intensely bright, and the calf seemed to glow.

"Luna," I said. "That's your name."

The little heifer just tucked her chin into her hind legs and drifted off to sleep. I hoped that Xena and the calf would sort everything out overnight and that in the morning I'd find a healthy calf, belly full of milk, glued to her momma.

The next day proved disheartening as I awoke to Xena's bellowing. She was lowing at her calf to follow her around the pasture. Little Luna did her best, but she wasn't even twenty-four hours old. She would follow Xena for a bit and then plop down and go to sleep. I didn't see her nurse, but I presumed she had, otherwise she'd have been a lot weaker.

That night, it began to rain. As usual, the herd migrated down to the loafing shed for shelter. Xena was the last to arrive, nudging her calf along. Ethel, though, chased them away.

I was beside myself. I tried to move the calf back to the shelter, but Xena wouldn't have it. She had received Ethel's message loud and clear. The calf nestled down in the tall pasture grass, shivering. The best I could do was cover her with hay.

By morning Xena and her calf were in the shelter lying down next to Ethel. Whatever the issue was, Ethel and Xena must have worked it out. Or so I thought. Over the next several days, Ethel exhibited odd behavior toward Luna. She would often hover over the calf and groom it. Poor Xena was smaller than Ethel and lower in the social hierarchy and couldn't do anything about it. Xena would stand there, head down, snorting in discontent while Ethel licked Luna.

Ethel was walking a fine line. She had her own calf to raise and yet she exhibited maternal behavior toward Xena's calf. Not to the point of nursing her, but enough to rattle Xena, who was already stressed from being a first-time mother. Sadly, cows do sometimes steal another's calf. There isn't much scientific research on this phenomenon, but the leading theory is a sort of spillover of maternal hormones in the older cows. This would make sense as Ethel had calved three weeks before and would have had higher levels of oxytocin than Lucy, who was six weeks out from calving. The only thing a farmer can do is ensure the cows have plenty of room to spread out so the first-calf heifer can separate herself from the others.

Xena did, in fact, learn to keep a distance from the herd. I rarely saw her nursing the calf during the day that first week, so I suspected that she was doing it under cover of darkness, when Ethel wouldn't bother them. By the end of the first week, Luna had put on a respectable seven pounds and was keeping up with the herd without any problem. Xena had become more comfortable with motherhood, allowing Luna to nurse during the daytime. Luna had also gotten her legs under her and began to spend more and more time with the other two calves. Xena always kept close watch on her, snorting disapproval if Luna got too far away.

Xena guarded her calf with her life. Early on, she had chased me away when I tried to move Luna to the shelter. As time went on, Xena became even more protective of her calf. When Luna was a week old, I picked her up for her final weighing, but Xena charged at me aggressively. She put her head down and rammed me repeatedly with her stubby horns until I put the calf down. Lucy and Ethel never did that. They might be unhappy when I picked up their calves, but they never attacked me. Xena's behavior, though unpleasant for me, was a normal response for a first-calf heifer. As long as I kept them together, and if Xena never had another calf, she would protect Luna until the day she died.

With three new calves, the herd stood at ten—a 42 percent increase in six weeks. The calves would be dependent on their mothers for at least three months, but by the fall of the third year I had a more complicated society of cows on my hands. There were two immediate impacts of the herd's growth. First, the dynamics changed. Previously, the cows had pretty much done everything together. They grazed together, slept together, and generally behaved as a unit. No doubt this was a hardwired instinct for protection.

But with ten cows, it wasn't physically possible for an individual to be in close proximity to everyone else. So the herd began to break into two groups. The fissuring was most apparent during grazing, as Couzin's models had predicted.

The second consequence of the enlarged herd dictated a change in my management strategy. First and foremost, everyone had to be fed. When the pastures were green, there was plenty of grass to eat, and I did not have to move the herd very frequently. The fall and winter, though, required careful attention to the quality of the forage. I needed more hay to get through the winter. With all those cows, I needed another hayrack and feed trough, just to give everyone room to eat.

The most interesting dynamics emerged during the evening playtime. Here, there was a clear tendency of the yearly cohorts to play together. Cricket, Walker, and Luna liked to chase each other, while Daisy and Tex paired off to butt heads. BB tended to associate with them, too, and I began calling them the Three Musketeers. (The calves were the Nachos.) Xena had become too uptight in her motherhood to join in, so she stayed mostly to herself, while the OGs would hang together. The calves sure seemed to enjoy these play sessions, and the amount of energy they expended playing was impressive. Surely play must serve some function other than just fun.

CHAPTER 16

Playtime

And surely all God's people, however serious and
savage, great or small, like to play. Whales and
elephants, dancing, humming gnats, and invisibly
small mischievous microbes—all are warm with
divine radium and must have lots of fun in them.

—John Muir, The Story of My Boyhood and Youth (1912)

M ost people's perceptions of cows are from the side of a
road, from where they appear to be inert objects of the
landscape, barely more animate than rocks. But at certain times
of day, especially in the evening, when my herd had finished grazing for the day and the heat was no longer oppressive, they kicked
up their heels and had some fun.

As is the nature of youth, the calves were always the first
to play. The sight of little calves—no bigger than deer fawns—
running doughnuts around the adults brought such peace to my
heart that it made all the hard work worth it. Ken would come over
many summer evenings and we'd crouch down to the calves' level
and play their little game. Every year it was the same. Tentative
at first, the calves would do little head waggles to get a bead on
whether these humans were friends or strangers. Then they'd
do little bunny hops and trot over to us. The bull calves were
braver than the heifers and would start head butting a knee.

If I put my hand out, they would push back with their foreheads. BB, Tex, then Walker all engaged in the bunting game. When the bull calves realized they couldn't win against a human, they usually went back to playing with the girls. Sometimes, though, they would take on one of the yearlings, or, if they were feeling particularly brave, they would try to rile Ricky Bobby. The girls would join in too. Cricket liked to tussle with Walker, and when Luna got big enough, she would often goad the others by giving a playful head bump, jumping up in the air like a bunny, and then running away.

While the juveniles were having fun, the adults would usually pair off and engage in mutual licking. Ricky Bobby, BB, and often Xena would come to me and jostle one another to be the first to put a head on my shoulder. Ricky Bobby, of course, always won.

Peace and joy flowed freely from the herd in these moments, like a fountain of youth. At sunset the sky turned salmon, and time ceased to mean anything, the only evidence of its existence heralded by nature's rhythm: deer foraging along the fence line, sometimes testing the cows' patience by entering the pasture; the bats chittering as they began their nightly exodus from the barn; and the lightning bugs flashing their secret code through the trees. Before I knew it, an hour would fly by, and Ken and I would find ourselves in darkness. We'd look to the heavens and marvel at Venus or Jupiter, low in the western sky.

With the tranquility of these playtimes, it was easy to be lulled into thinking that the cows played with each other just for the sheer joy of it. I had to remind myself that everything has a purpose—even play. And though the cows, especially the calves, took obvious pleasure in their games, that did not mean there wasn't some underlying biological reason for what they did. Quite the opposite. Some scientists have argued that every animal on the planet engages in some form of play, and because play is energetically expensive, it must serve an important biological function. The fact that play is fun makes it no less important. The enjoyment associated with play serves to reinforce the behavior so that an animal will continue to do it.

The scientific study of play has been plagued with difficulties, mostly because there hasn't been agreement on what constitutes play as opposed to some other type of motor activity. That said, I am drawn to the work of Gordon Burghardt, a professor of evolutionary biology at the University of Tennessee. Burghardt has studied creatures from reptiles to mammals and makes a compelling case that every species engages in some form of play activity.[1]

Before we can answer the question of why animals play, we have to define what play is. Burghardt identified three distinct types.

First, there is locomotor play. This is when a solitary animal engages in sustained movements like running and leaping without any obvious reason. One characteristic of locomotor play is direction reversal. The calves exemplified this type of activity by running across a stretch of pasture, only to reverse direction and scamper back to their mommas. This is usually the first type of play a juvenile engages in. The calves were typical, engaging in running bursts when they were a day old. Cricket started a mere four hours after she was born.

Head shaking and body twisting are also characteristic of locomotor play. Certainly, the calves did that too. Rarely did they just run. They did bunny hops, kicking up their hind legs while shaking their heads.

And it wasn't just the calves. The adults, especially the males, would, from time to time, do the same thing. It was quite a sight to see Ricky Bobby take off galloping across the pasture, tail held high with his switch streaming behind him like a flag. Inevitably, he would come stampeding back. Four hundred pounds of cattle at full tilt was something to respect, and I would get out of the way when the big boy got the zoomies.

The second type of play is object play. Anyone with a dog or a cat is familiar with object play in animals. Dogs love to play with balls, mostly as a game with humans, but even without a suitable partner, many dogs will play catch with themselves by tossing a ball into the air. Cats, the more solitary of the two species, are well-known for amusing themselves with objects, whether it is a ball of string, a piece of crumpled paper, or a paper bag.

Often these games mimic the predatory behavior of the animal, so it is no surprise that object play is more common in carnivores. Early on, I had put a yoga ball in the pasture. The cows sniffed it once and then ignored it. Never did they play with the ball. Ricky Bobby, however, was fond of pushing around heavy objects. He liked to tip over the mineral block—a fifty-pound cube of salt and other minerals. The cows were supposed to lick the block to get the minerals the grass didn't provide, but Ricky Bobby liked to move it around. He also liked to manipulate the hayrack. It became such a nuisance to find the rack tipped over and hay scattered on the ground that eventually

I had to stake it in place. He did the same with the feed trough, flipping it end over end until he had wedged it against the fence. Sometimes he even pushed around the water trough, which weighed three hundred pounds when full.

The third type of play is social play. Unlike locomotor or object play, which are solitary activities, social play is always with another animal (*conspecifics*, as scientists like to say). With dogs, for example, social play usually takes the form of chasing each other or engaging in play fighting, where dogs take turns pinning one another to the ground. Critically, social play is reciprocal, where the animals take turns chasing or pinning each other. And when one animal is bigger than the other, the larger one will self-handicap and go easy on the little guy.

Some of these games, especially in predatory species, may look to human eyes as if they are verging on aggression. Dogs may growl and nip at each other. In order to avoid escalation, all animals have universally recognized play signals that communicate that it is just a game. In the 1970s, Mark Bekoff, an animal behavior scientist at the University of Colorado, pioneered the study of play behavior in canids. Bekoff observed that dogs communicated that no aggression is intended through a variety of signals, but most commonly with the play bow, first documented by Charles Darwin.[2] Primates don't bow. Instead, they make a play face—a relaxed open mouth with upturned corners sometimes accompanied by panting or laughing, easily recognized in both chimpanzees and humans.[3] Play signals, although they differ between species, have the quality of being unambiguous. They have to be. If there were any ambiguity, then the play signal might be misinterpreted as an aggression display.

Interestingly, there hasn't been much attention given to play signals in cattle. This is curious as it has been well documented that cattle play. As just noted, there must be unambiguous signals that cows exchange so they know when they're engaging in a game. One of the few studies that mentioned a play signal (although the authors didn't call it that) described three types of games that calves played: mounting, pushing, and head butting. The authors noted that the pushing game was often instigated "with an initial jerk of the head and/or bounce of the body."[4]

In search of the cow play signal, I started paying close attention to the new calves: Cricket, Walker, and Luna. I could count on one of them to instigate a game with someone else every evening. At first, it seemed chaotic. One

Cowpuppy

calf would head butt another, and then they would chase each other around the corral or out into the pasture.

But very quickly it became obvious that there were, in fact, signals being exchanged. Although cows can bow (they do it when they go down on their forelimbs as the first step in lying down), they don't do a full bow when playing. The instigator puts the head down and extends the neck in line with the back. Although the head is down, the chin is not tucked in. Much of the power of a cow comes from those neck muscles, which is maximal from the tucked position. From the extended position, little harm can be done. In contrast, a head-down, chin-tucked position places the horns at an aggressive angle. Playful intent is also signaled by a quick side-to-side wag of the head.

The other component of the play signal is the hind end. Almost like a dog, a playful calf holds its butt high in the air, often giving a little kick of the hind legs. They also wag their tails when they are having fun (and when they are nursing). Because a cow's power also comes from those massive legs, when they are off the ground, they effectively handicap themselves. This posture is clearly distinguished from that of an angry bull: head tucked in, horns forward, back arched in an inverted C, rear legs coiled and ready to charge.

Sometimes play took unexpected forms. What really got everyone excited, though, was a pile of sand. I had had a dump truck drop a load of sand in the corral with the intention of spreading it around the areas where the cows congregated—under the oak tree with the water trough and hayrack as well as the loafing shed. But before I could distribute the sand, the cows went wild over it. Even Lucy and Ethel, whom I had never seen so excited, took part in the games. They played king of the mountain, taking turns standing at the peak and running the others off. They kneeled down on their front legs and rubbed their horns in the sand, only to take off running and jumping, flinging the sand this way and that.

One of the characteristics that distinguishes play from other types of activity is that there is no other goal than play itself. This is important because some of the things the calves did had different purposes in different contexts. Mostly I'm talking about mounting behavior. It would often start with one calf putting its head on the other's back, followed by rising up and straddling the rump. The mounter would usually make a funny little grunt, and the mountee would immediately run away, sometimes to return and reverse roles. Beginning when they were about four weeks old, both the boys and girls did it, although the boys did more of the mounting. Obviously this is the same behavior that adult cows engage in when they mate—including the females—but as a play activity among calves there was no reproductive intent. The game itself was the goal. Sometimes, when I crouched down with my back to them, BB and, later, Tex, would try to mount me. I could usually tell when they were fixin' to do it by the way they held their head or the little grunt as they started to rear up. I didn't really care to be mounted, so I would just stand up. In fairness to the calves, I would often lay my head on their withers and give them bear hugs, which they might have perceived as being mounted. In their minds, they were just doing the same to me.

Given that play is energetically expensive, why do animals do it at all? Most animals live on the edge of survival. Play would seem to be a frivolous waste of calories. Precisely because play is so costly, it must serve an important function in terms of an animal's development and survival. Herbert Spencer, a nineteenth-century psychologist, formalized one of the earliest theories of play by suggesting it simply burned off surplus energy.[5] Because play had no other end goal, Spencer argued that it was not essential for life. Only "higher" animals could afford to engage in it because their nervous systems were more efficient.

A century later, as play was documented in more and more species, including so-called lower animals, like fish, Spencer's theory fell out of favor. It is worth noting that when animals are sick or malnourished, they don't play, suggesting that play, however important it is, still takes a back seat to basic survival functions. Conversely, if an animal is not engaging in age-appropriate play, then it could be a sign that something is physically wrong.

A few years after Spencer published his theory, William Lindsay, a Scottish physician interested in afflictions of the mind, wrote an influential treatise on *Mind in the Lower Animals*.[6] Lindsay described animal instincts—drives that animals are born with. He had a rather lengthy list of twenty instincts, ranging from the physical (such as thirst and hunger), to the sexual and social, to the sense of supernatural agency, which formed the basis of religiosity. Yes, Lindsay believed other animals were capable of religious experiences, pointing to the common practice of dogs accompanying people to church and even having spots in the pews reserved for them. Lindsay also thought that play was an instinct, especially in the young. The play instinct lets young animals master critical skills and assess their abilities. His instinct-practice theory is currently one of the leading theories of play.[7] Given the similarity of the games the calves played to "adult" activities, this makes a lot of sense. However, it didn't explain why the adults still played. After all, they had already mastered these skills.

There is no simple answer to why animals play. The current consensus is that there are multiple reasons, a framework originated by Harvey Carr, a comparative psychologist at the University of Chicago, in the early twentieth century.[8] Carr listed seven reasons for play: diversion, catharsis, alleviation, recuperation, practice, education (which included both exercise and instinct-practice), and social bonding. That he listed diversion as the first reason is important because he recognized that animals engage in some activities for the pure joy of it. Certainly that was the impression the calves gave when they gamboled about the pasture. They weren't conscious of any other reason for their activities other than that it was fun.

The exercise component of play is worth unpacking. Carr was referring specifically to the development of perceptual-motor skills—what humans call *hand-eye coordination*. W. A. D. Brownlee, a British veterinarian, fully articulated the theory in 1954 to explain the behavior of dairy cows.[9] Brownlee thought that the play calves engaged in aided the development of muscles that were used for vital activities like fighting and fleeing. Without such early development the muscles would atrophy. The exercise theory was later broadened to encompass the nervous system itself. Recognizing that the brain goes through critical periods of development when skills must be acquired, play becomes critical for syncing up the brain and muscles.[10] While Brownlee thought play was mostly about muscle development, we now recognize that play also develops the senses: sight, sound, smell, and proprioception, also known as body awareness. BB seemed to be a bit deficient in this last domain, for he was always stepping in his poop while everyone else had no trouble avoiding such mishaps. Although not blessed with perfect coordination, BB had plenty of other charms.

Finally, nonsolitary play should, in theory, teach individuals how to get along with one another. While this may be true for human children, the socialization function of play has been harder to prove for other animals. For example, juveniles of solitary species, like bears, play just as much as social species.[11] Sex differences, though, seem pervasive. Males engage and initiate play fighting more than females. The extent of these differences is greatest in polygynous species where males fight for the reproductive rights of a group of females. (Hyenas are an exception. Females are dominant over males, and juvenile females initiate play more than males.)

What about cows? They are highly social, and under human management, a herd would have only one or two bulls who are kept separate from the females except when the farmer wants them bred. After a few seasons the bull would be exchanged for another to maintain genetic diversity in the herd. This separation mirrors how their wild ungulate cousins—bison, antelope, deer— live. The sexes stay segregated except during the mating season (the rut). The separation of sexes is an evolved evolutionary strategy in which males maximize their body condition before the rut while females protect the prior year's offspring as long as possible.[12] Consistent with the exercise hypothesis, the juvenile males have to develop skills that will allow them to run with the big boys. This is why the male calves—BB, Tex, and Walker—were generally more avid in initiating games than the females. Boys will be boys.

I think precisely because of the males' inborn tendencies to be more outgoing and playful as calves, it was inevitable that I formed a close relationship with one of them. Maybe it was because he was the first, or maybe it was just the luck of the genetic lottery, but BB and I formed a fast bond. He was always the most gregarious of the bunch. Even after three seasons of calving, he was still my favorite. Ricky Bobby had his charms too. When he played, it wasn't as human-friendly as with BB. Ricky Bobby was more like a needy man-baby, insisting on brisket rubs by placing his head on my shoulder. Tex inherited some of Lucy's stoicism. He enjoyed playing with Daisy, but he didn't initiate playtime very often.

None of this is to say that the females were unfriendly or unplayful. But by the time they were yearlings, they had settled into the calm reserve characteristic of cows. Daisy was a cool cucumber, quite unflappable, while Xena was always a bit unpredictable. Much to my surprise, Lucy became very demonstrative with me, lifting her head for long neck scratches. Apart from the sand pile, she was never interested in playing, but her stoic demeanor had a calming effect on anyone she let into her personal space. Lucy also turned out to be exceptionally taken with her own appearance, which led me to a unique experiment.

CHAPTER 17

Mirror, Mirror

To live without mirrors is to live without the self.

–Margaret Atwood, "Marrying the Hangman" (1987)

There is perhaps no more storied test of consciousness in the animal kingdom than the mirror test. As Margaret Atwood put it so succinctly, mirrors allow an individual to see an image of themselves. But here's the million-dollar question: When confronted with one's reflection, does the individual actually recognize it as themselves? To do so, they need to have self-awareness, a sense of "me." But what constitutes self-recognition in the mirror? How does an animal—or a human infant, for that matter—demonstrate they recognize their own reflection?

Early studies were notoriously subjective, citing unusual behavior toward a mirror. Many of these methodological issues have been worked out, and legitimate claims of self-recognition have been made for a few species.[1] Cows, though, were not one of them.

The basic problem is that when you put a mirror in front of animals, they react in all sorts of peculiar ways. Birds have long been known to attack their reflections. If you want to elicit a showy display of color in a male Siamese fighting fish, place a mirror next to his tank. He will act like his reflection is another male. Dogs will usually ignore mirrors, but sometimes they bark at their reflections.

Some of these reactions may stem from a mirror's novelty. Unlike humans, who have access to mirrors from an early age, animals rarely encounter mirrors in the natural world. There are reflective surfaces, such as water, but they are not usually flat enough to produce a high-quality image. In urban environments, windowpanes can serve as mirrors, but only animals who live in those areas have the opportunity to gain experience with their reflections. Cows would not fall in that category. In fact, I could find no reference to cows and mirrors—either in the research literature or on the internet.

The classic mirror test was developed in the 1960s by Gordon Gallup, a psychologist at Tulane University, and remains the gold standard of self-recognition.[2] Gallup placed a full-length mirror in the cage of two male and two female juvenile chimpanzees. For the next ten days, he recorded the behaviors the chimps made toward their reflections. At first, the chimps made frequent social behaviors toward the mirror, including bobbing up and down, vocalizing, and making threatening displays. By the third day, however, these actions were replaced with self-directed behaviors like making faces at themselves, blowing bubbles, and watching themselves manipulate wads of food. To test if the chimps had achieved self-recognition, Gallup anesthetized the animals and placed a red mark above one of the eyebrows of each chimp. They were then allowed to wake up in their cage. Upon seeing themselves in the mirror, the chimps reached up and touched the marks on their heads. Sometimes they touched the mark and then examined their fingers, even though the red dye was completely dry. As a control condition, Gallup did the same procedure to two chimps that hadn't had prior mirror exposure. When they awoke in their cage, they acted toward their reflections like the other chimps had done the first time they encountered a mirror. Critically, though, the naive chimps didn't probe the marks on their heads, proving that the original group had learned to recognize themselves over ten days of exposure. When Gallup tested monkeys, none of them showed any self-directed behavior or mark-recognition. They continued to behave as if their reflections were other monkeys.[3]

Although versions of the mirror test have been tried on many species, all except the aforementioned primates have failed Gallup's strict criteria. Some scientists have argued that the mirror test is biased toward visual species and that allowances should be made for species that rely on other senses, like an olfactory version for dogs.[4] The field is rather split in this regard. Some researchers hew to Gallup's original test as an all-or-nothing gold standard for self-recognition, while others favor a continuum of self-recognition that allows for different modalities.[5] Even if an animal does not meet Gallup's criteria of self-recognition, it may still use a mirror in other ways. Self-directed behavior is really the highest level of mirror usage.

Short of exploring a mark on the head, animals that pay attention to a mirror go through several levels of exploration.[6] The first-level mirror response is acting like the image is another animal. Aggressive displays and vocalizations are characteristic. The second level is physical inspection, like smelling or looking behind the mirror. The third level is called *contingency checking* and is characterized by repetitive behaviors, like head bobbing or assuming postures that afford views of themselves. And the final level is self-exploration of marks on the head and body.

Although there was no documentation of a cow mirror test, it had been attempted in a few other farm animals, notably horses, sheep, pigs, and even roosters. Sheep exhibited first- and second-level behaviors like staring at their images, licking them, and vocalizing. Some individuals demonstrated contingency checking by moving their heads in relation to their reflection or tracking other individuals in the mirror.[7] None of them could use the mirror to navigate to a treat in a maze, however.

In a study of horses, they, too, showed first- and second-level behaviors, including prolonged staring, looking behind the mirror, licking, and, interestingly, opening the mouth and protruding the tongue—a level-three contingency checking behavior.[8] The researchers then used gel to place an X on one of the horses' cheeks. In one condition the gel was transparent, and in another condition the X was colored with either blue or yellow dye. The transparent gel controlled for the possibility that the horses would react to the physical presence of a gel. The horses reacted by trying to scrape off both the transparent and colored marks. Although some of the horses displayed more scraping behaviors with the colored marks, it wasn't consistent, and because only four horses were tested, the researchers couldn't reach any statistical conclusions.

Pigs were intermediate between sheep and horses. In one study, pigs exposed to a mirror reacted with first- and second-level behaviors. (One even broke the mirror charging at her image.)[9] Pigs with mirror experience were able to use the reflections to find a food bowl, whereas mirror-naive pigs took longer. Roosters also appear able to recognize that their reflections are not other roosters because they don't vocalize alarm calls to their reflections.[10]

The only study of cows and mirrors wasn't a mirror test per se. The question was whether a mirror would lower the stress of isolation while cows were being weighed. It had a modest effect of decreasing heart rate if the mirror was face-on, but no effect with a mirror positioned for a side view.[11] This raises the question of whether cows would behave more like horses or sheep in a true mirror test of self-recognition.

To prepare the cows for the mirror, I lashed a sheet of plywood to the fence. This would acclimate the cows to its presence, so any reactions they had to the mirror would be from their reflections and not the introduction of a new object to the pasture. The cow cam showed they barely paid the plywood any attention the first day. After that, they ignored it entirely. I would have put the mirror up sooner, but I wanted to wait until all the calves were born. So after the plywood had been up for a month, I swapped it out for the mirror. To avoid anything breakable, I glued two sheets of acrylic mirror to plywood. Two-by-fours as backing flattened the sheet, mitigating the circus funhouse effect.

Luna was the first to approach the mirror. She strolled by nonchalantly and then, catching sight of her reflection, changed course and went up to the mirror. She sniffed then looked behind it to find this new calf. She did a head bob and a little bunny hop, indicating she wanted to play. Her reflection did the same. She put her head down and leaned her forehead against the reflection, playing the head-butting game that all calves do. It must not have been as fun as playing with a real calf because she quickly lost interest and wandered away.

After ten minutes, the rest of the herd caught sight of their reflections and approached tentatively, maintaining a safe distance of twenty feet. Tex was the most exploratory of the bunch, creeping up to the mirror for a sniff. With all the cows cramming into the field of view, anytime one of them moved, it spooked the others, and the herd would stampede away. Lucy, Ethel, and Xena stayed far away, protecting their calves.

The herd spent no more than twenty minutes investigating the mirror that first day. But even so, their responses made two important things clear.

First, they saw their reflections. The bovine visual system didn't evolve in an environment with mirrors, so it wasn't a foregone conclusion that the cows would be able to focus on their reflections.

Mirrors are funny things. To perceive a reflection, you have to focus on the image *behind* the mirror, as if it is a piece of glass. The cows might have focused on the plane of the mirror itself, which would have rendered their reflections blurry. Their reactions made it clear they were focusing on their images. Second, they recognized these images as cows—although not necessarily as representations of themselves. They were confused by these ersatz bovines, but if they had thought they were anything other than cows, they would have run away immediately. So right away, it was apparent that the herd displayed first- and second-level behaviors.

On the second day, everyone in the herd took time to stand in front of the mirror. In fact, they spent so much time staring at themselves I started to become concerned that they wouldn't graze. This turned out not to be a problem, as they had twenty-minute bouts of investigation in the morning, afternoon, and evening but otherwise carried on as usual.

Each of the cows had a slightly different reaction to the mirror. Tex initially spent the most time staring at his reflection. He would stand there, swaying back and forth, processing the self-induced movement. Then he would stick his head through the fence and look behind the mirror. Daisy mooed softly at her reflection. Ethel, always the most vocal, mooed loudly at hers.

Ricky Bobby had the most interesting reaction. True to his cautious nature, he first stood behind the other cows, resting his head on one of their backs, and peered at his reflection using his herd mates as a shield. By the end of the second day, though, I found Ethel and him in front of the mirror. Ricky Bobby held his head low while Ethel groomed him, all the while watching what was happening in the mirror. He was rapt with attention, like a customer watching his hair get cut at the barbershop.

By the end of the second day, the mirror was smeared with nose prints, suggesting that the cows did not recognize the reflections as their own. It was likely that they were trying to scent these mysterious cows. I would not have expected them to recognize their reflections straightaway. After all, children don't recognize themselves until they are eighteen to twenty-four months old. The question was whether the cows would learn to recognize their reflections given the opportunity to stand in front of the mirror as much as they wanted. Given the rarity of this ability in the animal kingdom, odds were against it. But time would tell.

After a month, the cows had mostly habituated to the presence of the mirror. They didn't spend much time gazing at their reflections or looking behind the mirror, as they did the first week. Ricky Bobby was the last hold-out. I would still catch him investigating the mirror or, if he was feeling cantankerous, pushing the reflection with his head. Even after a month, though, I didn't see any evidence of repetitive movements that would indicate third-level understanding.

BB did, however, use the mirror in ways that suggested a higher-level understanding. I had positioned a trail cam next to the mirror to record what was happening when I wasn't there. One evening, two months after the mirror had been installed, the recordings showed BB standing in front of the mirror watching himself chew grass. He watched his reflection for a bit and then looked behind the mirror before returning to watch his reflection. Suddenly he turned his head to look behind him again and then moved aside. A few seconds later Lucy appeared and took up position in front of the mirror. The orientation of BB's head indicated that he had seen Lucy's reflection as she approached from behind and that he used that information to turn around.

Two days later, the roles were reversed. Lucy took up position at an angle to the mirror. The rest of the herd was grazing about twenty yards away. From her vantage point she alternately looked at their reflections in the mirror and then at them. BB stared right back at her. After a few minutes he sauntered over to the mirror, but Lucy chased him off. Ricky Bobby joined the scrum, too, but all he did was head butt his reflection.

These were remarkable developments in the cows' understanding of how the mirror worked. I was witnessing the transition from looking behind the mirror to looking at the source of the reflections. This approached third-level mirror behavior.

After four months, I figured the cows had had plenty of experience with the mirror. They still interacted with it on a daily basis, spending a few moments looking at their reflections as well as the others'. Ricky Bobby never seemed to figure it out, though, and preferred to bash his head against the plexiglass. Lucy spent the most time in front of the mirror, and she, more than the others, seemed to grasp the concept of the reflections. So she was a natural choice to place a mark on her head for the final, fourth-level test. I would mark BB, too, because his fur was so white and also because he was the first to have shown interest in the other cows' reflections.

I placed a splotch of blue body paint on Lucy's left cheek and above her left eye. Same for BB. Later that day, the trail cam recorded Lucy staring at her reflection. She turned her head a little to the right, giving her a better view of the mark on the left side of her head. As she was already the one who spent the most time in front of the mirror, I couldn't tell if this was just her usual level of interest or whether she had recognized that something had changed. BB didn't show any extra interest in his reflection that day. Two days later, I repeated the experiment, this time including Tex and also moving the mark to the left withers. Again, only Lucy took up a posture that afforded a view of the mark. Was it a coincidence? I couldn't say.

Even if the cows didn't reach the highest level of mirror recognition, their behavior showed an evolution from looking behind the mirror to using the reflections. It took several months, which was far longer than most experimenters had given their animal subjects, but it showed that cows were capable of learning something about mirrors. Their understanding of the world was not static; it was able to adapt to changes in their environment, possibly to the point that would qualify as some form of self-recognition.

After the first week, none of them ever mooed at their reflection again. Many dogs bark at their reflections, suggesting they think it's another dog. The fact that the cows stopped vocalizing at the mirror implied that they learned the reflections weren't other cows. But then again, they only vocalized under specific conditions. To fully appreciate the lack of vocalization, we must take a deep dive into cow communication.

CHAPTER 18
What Does the Cow Say?

With a moo-moo here and a moo-moo there,
Here a moo,
There a moo,
Everywhere a moo-moo.

–"Old MacDonald Had a Farm"

Everyone knows what a cow sounds like. The cow says *Moooo*! A classic example of onomatopoeia, a word that sounds like its meaning.

Or is it? In English, the cow may say *moo*, but in French, they *meuh*. In Italian, Spanish, and Portuguese it's *mu*. Welsh cows say the unpronounceable *mw*. Danish cows go *muh*, but in Dutch they *boe*. While most of the world seems to agree on the monosyllabic /m/ sound, to my ear it wasn't really an /m/ sound, or even one syllable. If we're going for acoustic verisimilitude, a few languages get pretty close to cow-talk. In Finnish, it's *ammuu*. In Bengali, it's *hamba*. But I think the closest to what cows really sound like, especially with emphasis on the second syllable, is heard in Tagalog: *ungaa*; or Korean: *eum mae*.[1] Of course, my cows were of the Asian variety, so that may have had something to do with the similarity of their vocalizations to languages of southeast Asia—a cattle dialect of sorts.

There is, in fact, tremendous variation in the sounds

167

cows make. Each cow has a repertoire of vocalizations that presumably mean different things, and no cow sounds exactly like another. As in the human voice system, cow vocalizations originate with vibration of the vocal folds of the larynx. The emerging sounds are then modulated by the configuration of the tongue, mouth, and lips. Sometimes a cow will vocalize with her lips closed, in which case it really does sound like *mooo*. Other times she will keep her mouth open, resulting in a nasal version, like Steve Urkel might do if he were a cow. And then there's *lowing*, a deeper, more resonant sound that can be heard at great distance.

It is a generally accepted tenet of biology that the reason animals vocalize is to communicate with one another. But for this to work, there has to be both a sender and a receiver. Although humans talk to themselves, other animals don't. Even the solitary wolf, seemingly howling at the moon, is communicating with other wolves, who may be miles away. He might be calling for a mate or warning others to stay away from his territory. Vocalization is intentional and meant to convey information to a recipient.

To understand what cows are saying, we have to examine mooing from both the sender's and receiver's perspective. What is the sender trying to convey? And what do they expect the receiver to do? If we understood the receiving end, this would open the door to verbal communication, like people have with dogs.

All of my cows mooed, some more than others, and the females were generally more vocal than the males. It wasn't until the third season that I could say this with some confidence because it wasn't until then that I had enough males and females to observe. The boys were more physical, but the girls were more vocal. A bit like humans, I guess.

Ethel had always been the most vocal of the herd. She mooed nearly every evening when it was time for the grain. She mooed incessantly at each of her calves until they were a month old, to the point that it appeared they learned to ignore her entreaties. Sometimes I would hear her mooing in the pasture for reasons only she knew. Unless it was repetitive or sounded particularly urgent, I ignored her, just like the rest of the herd did. (Okay, I didn't really know if the herd ignored her, but there were no obvious reactions from the other cows when Ethel was doing her regular mooing.)

Although Ethel was the most vocal of the herd, the others mooed from time to time, and for a few weeks after Xena's calf was born, she mooed even more than Ethel. As a first-time mom, Xena did not possess the knowledge and security of a seasoned cow, like her momma, Lucy. Newborn calves don't do much other than drink milk and sleep. The momma, though, has an increased caloric requirement due to lactation and needs to eat roughly 50 percent more than normal. Poor Xena was torn between filling her belly with grass and checking on her calf. She would graze for a bit with Luna sleeping nearby, but as soon as Xena moved on to fresh forage she would bellow at Luna until she got up and followed her to the new location. Ethel had done the same with all of her calves, but Lucy radiated confidence and rarely needed to moo. Her calves just followed her.

Except when a calf managed to slip through the fence. It was inevitable. Every one of the calves, at some point, found their way to the wrong side of the fence. They had a tendency to nap beneath the rails while their mommas were grazing, and I think when they woke up, they just rolled the wrong way. Other times, they were curious and squeezed underneath. However they got out, though, the momma went into a tizzy, running back and forth along the fence line, mooing desperately at the wayward calf to get back in, which, of course, they never seemed to figure out. One time, when Tex was a calf, we found him curled up under a tree in the forest. When Lucy mooed, I knew something was up. The only clue to his approximate location was the direction Lucy was aiming her bellows.

The boys' vocalizations were entirely different than the girls'. They rarely mooed, and when they did, it was a harsh sound, almost like a donkey braying. The only times Ricky Bobby mooed were when he was separated from the herd. This didn't happen often, just when he was castrated and when he was having his hooves trimmed. I think his braying was more from the pain of separation than the physical pain. BB's moo sounded more like a grunt, but he, too, did it only when he inadvertently got separated from the herd, like when he lingered in the barn too long and the herd decided to leave. Poor BB would come out and, realizing he was alone, call out for his friends.

These descriptions would seem to imply that cows moo mainly when they're in distress. Temple Grandin used cow vocalizations as a type of negative metric of how well they were being handled at slaughter.[2] Vocalization is also increased when calves are separated from their mothers (more so with the mothers than the calves).[3] The same is true for dairy cattle.[4] For these reasons much of the cattle industry has gravitated toward mooing as a marker of stress.[5] A quiet cow is a happy cow. However, the context of any vocalization must be taken into account. Slaughterhouses and dairies are inherently stressful situations, so any vocalization in those contexts will necessarily indicate physical or emotional stress. What about other, non-stressful situations?

Although my cows vocalized most prominently when they were unhappy about something, they also mooed when they appeared perfectly fine. Like Ethel, they occasionally mooed in excitement when I was about to give them treats or when I was moving them to a fresh pasture. When the mommas mooed at their calves, sometimes they were just saying, *Hey! Let's go!* Other times, I think they were teaching the calves what their momma sounded like. Cows identify each other by their signature calls. One study found that three- to five-week-old calves can differentiate recordings of their mothers from other cows.[6] And it's not just the cows identifying one another. I could do it, too, just as I could identify each of my dogs by how they sounded.

In an effort to understand cattle vocalizations, researchers have put considerable effort into decoding moos. The classic approach was through acoustic analysis, which relied on a spectrogram. A sort of vocal fingerprint, a spectrogram plots how different audio frequencies change over the course of a vocalization. They are especially common in the analysis of birdsong. When applied to speech, acoustic analysis is termed *phonetics*. The fundamental unit of phonetics is a *formant*, which describes the frequency of a sound. Speech, though, is composed of several frequencies, each changing in amplitude during the course of a vocalization. The formant with the lowest frequency is called *F1*, the second *F2*, and so on. When this type of analysis was done on cow vocalizations, the average F1 was 790 Hz and F2 was 1942 Hz.[7] This combination of F1 and F2 sounds similar to the formants produced when an adult human male makes the long *a* sound.[8]

While formants describe vowel sounds, speech involves a lot more than that. How a formant is initiated and how it ends determines the basic unit of human speech, called a *phoneme*. Even though there are five vowels in written English, there are twenty to twenty-four phonemes that describe how to pronounce vowel sounds. For example, in grade school, kids are taught the difference between the short *a* sound in *cat* (abbreviated /a/) and the long *a* of *cake* (/æ/). There are another twenty or so phonemes for the consonants. Cattle vocabulary, if there is one, would have to use a much more restricted set of phonemes because cows don't have nearly the degree of vocal control humans do.

A single cow vocalization lasts one to one and a half seconds, but there are several ways a cow can alter how she moos. Presumably, these alterations are done intentionally and, therefore, contain information that is being transmitted to someone—another cow, a human, or maybe another farm animal. One study found that moos could be divided into open- and closed-mouth types. With mouth closed, a cow produces a lower frequency formant, and these types of moos are more frequent when a momma is near her calf, especially during the first month after birth.[9] With mouth open, the sound projects farther and is a higher pitch. These types of moos are used when momma and calf are separated and the momma is calling for her calf. Further analysis has revealed that individual cows could be distinguished by differences in the formant frequencies—a vocal signature—but only in the closed-mouth moos. The authors speculated that vocal identification is more important when cows and calves are near each other, with open-mouth moos serving as a more general call. Taking this type of analysis one step further, Australian researchers determined that vocal signatures are maintained across different emotional states.[10] Cows were assumed to be in a positive emotional state when the farmer was getting their feed ready and a negative state when they were isolated or denied access to food, but the cows were recognizable by their moos in both positive and negative contexts.

No study has yet been able to differentiate positive and negative emotional states just from mooing.[11] It may be that cows, like most animals, have no need of mooing when they're happy, instead reserving vocalization for distress and warning calls. But that wouldn't explain Ethel's moos during the evening grain. No, vocal decoding is a hard problem. Machine-learning algorithms are just beginning to get a handle on decoding human emotion from human speech.[12]

The difficulty is compounded in domesticated species. Because humans have altered their natural evolution, the vocalizations of domesticated animals may be directed at either their compatriots or humans. Wild animals, in contrast, have no need to communicate with humans, so their vocalizations are directed at one another or occasionally to frighten off potential threats.

In terms of domesticated animals, the most progress in understanding vocalizations is with dogs. Dogs are evolved from wolves, which are famously vocal, and this may have something to do with dogs' ability to understand human speech, as well as our ability to decipher their vocalizations. Research has shown that people are generally quite good at recognizing some dog emotions from their vocalizations, such as aggressive barks at strangers versus barks when they're playing, although this ability is largely dependent on a given person's experience with and thus exposure to dogs.[13] In general, low-pitched barks are understood to be aggressive in nature, whereas high-pitched barks convey a more positive emotional state, like playfulness.[14]

In fact, this high-low vocal rule has been observed in many animals. Harsh, low pitches signal hostile intent, and purer, high pitches mean friendliness.[15] There is a simple explanation for this relationship. Physics dictates that large animals will have lower-pitched vocalizations. To the extent that an individual can vary the pitch of its sounds, lowering the pitch will make them sound bigger and more threatening. Conversely, raising the pitch will have the opposite effect, signaling that they are small—even juvenile—and don't want any conflict. Although this rule has not been studied with cattle, it likely holds true.

If humans can broadly recognize whether an animal vocalization is hostile or friendly, it is straightforward to program a computer to do the same. It's really a matter of having a lot of data. An early decoder of dog barks, for example, could tell the difference between barks produced in six situations: playing, fighting, walking, being left alone, being approached by a stranger, and being shown a ball.[16] However, the algorithm required six thousand barks and was about 50 percent accurate (substantially better than guessing, which would be correct 17 percent of the time). To make a moo-alyzer, you would need a comparable number of audio samples, about one thousand per type of moo. Doable, but a person would still need to hand code what each moo represented before training a neural net on the data. One promising approach to decoding cow vocalizations recognizes that moos often occur in sequence, almost like a sentence. Instead of focusing on a single moo, the transitions between open- and closed-mouth mooing suggest a simple way to understand what a cow is saying, especially in the context of mother-calf interactions. In a typical sequence, the momma uses an open-mouth moo to call her calf and then transitions to closed-mouth when they are reunited.[17]

Mooing isn't even the whole story of cow communication. Cows make other types of vocalizations more frequently than moos. They grunt and

  Cowpuppy

snort. Ethel, as the most vocal of the herd, was a prolific snorter too. She would only do this when her calves were doing something she didn't like, usually by wandering off too much for her comfort. What was interesting was that Ethel used a graded level of snorting. Mild irritation was signaled by a soft, brief snort. As she became more unhappy, this would increase in volume and stridor to the point of sounding like a grunt. The best way I can describe it is how Marge Simpson sounds when she grunts in disapproval. Although scientists have studied cow mooing, there has been no research on these types of subvocalizations in cattle. However, grunts have been categorized in other ungulates. Deer have been said to emit four types of grunts/snorts distinguishable by their acoustic parameters: the grunt, the alert-snort, the snort-wheeze, and the aggressive snort. All but the alert-snort are produced when deer are near each other, some being affiliative and others aggressive. The alert-snort is a loud propulsion of air by a lone deer warning others within earshot of a potential threat.[18] I heard alert-snorts every day when I walked through the woods around the farm. It seems likely that cows have a similar vocabulary. I would often hear the cows emit brief, high-pitched grunts when they played with each other. This was especially true of the calves, but everyone did it when they got going, like when they played in the sand pile.

There was another type of cow vocalization that I could find no reference to. I call it *cooing*. This was something all the calves did. When I crouched down and called them over, they would often nuzzle my neck, licking it while making soft snuffling sounds, which sounded like the coos of an infant. It was so quiet, you could only hear it if you were on the receiving end. I assumed the calves did it with their mommas too. Sometimes even the adults cooed. Xena tended to do it a lot when she wanted neck scratches. The coo of human babies is a potent releaser of oxytocin in mothers, and scientists think this promotes and maintains bonds with the parents. I think cows do the same thing, which would explain why I became so attached to them, especially the calves.

Despite the research on cattle vocalizations, there has been none on their ability to learn human vocal commands. Can cows understand human language? Based on my observations, the answer appears to be a qualified yes.

Take their names, for example. There would be many times when BB was walking along with the other cows, minding his own business, and I would call out, "BB!" He usually turned his head and looked straight at me. He didn't do it if I called one of the other cows. Sometimes he would trot over to see what was up. Xena was more subtle. She rarely looked toward me when I called her name, but she would often orient an ear in my general direction, indicating that something registered. Ricky Bobby did the same, but neither Lucy nor Ethel did. Tex and Daisy were ambiguous, sometimes showing signs of recognition, while other times they ignored me. Lucy and Ethel were the oldest of the herd, so it may be there is a critical period during calfhood for rudimentary language acquisition, just as with human children.

As the herd grew in number, it became increasingly difficult to call a single cow. They were always together, so when I called for BB there was no way for the other cows to know that I was singling him out. BB, of course, was my favorite, so he got more than his fair share of attention, which I'm sure contributed to his knowing his name. I did my best to talk to each cow every day, speaking each one's name softly into their ears so only they could hear it.

A name is a powerful thing. Humans name something when they need to denote its uniqueness. Livestock do not have names because it is too hard to kill a named animal. Many farmers, though, will name a few special animals. Maybe it is a dairy cow that has been especially productive, a charismatic pig, or a bottle-fed calf that becomes a child's pet. Some might argue that naming an animal is an act of dominion over it.[19] But I think it goes beyond that. An animal with a name has made the leap from *it* to *he* or *she*. Named animals are no longer objects.

What the cows made of their names was an entirely different question. Although there has been no scientific study of cows' perception of their names, we can look to research on dogs for a clue. How do dogs learn their names? Through eye contact. Scientists call this an *ostensive signal*, which is a body cue that precedes or is concurrent with a communication signal. When a human looks at a dog and speaks its name, the dog makes an association between hearing its name and paying attention.[20] Cows are exquisitely sensitive to head direction, and they can tell when you're looking at them.

There is no reason why the same trick wouldn't work with them too. It did with BB and, to varying degrees, the other cows. But what I found worked better than simply looking at the cows was making a movement with my hand that motioned them to "come here." Speaking their names at the same time helped them make the association.

Even if farmers don't name their cows, they still need to communicate with them. The most basic task is to summon the herd. Whether to feed them, milk them, or perform routine veterinary care, a cattleman has to call his herd up from the pasture. The number of ways this can be done is as varied as the farmers themselves. Some will whistle. Some will call out, "Hey, cows!" Some will use the sound of the tractor or the rattling of a can. Anything the cows associate with food will work. If the herd has a large range, though, they might not hear the farmer. In Sweden, they have *kulning*, the ancient art of singing to cows, which can be heard over long distances. In the Alps, they yodel. While perhaps not as lyrical as *kulning*, in the United States, we have hollerin', which is a countrified version of yodeling and serves the same purpose. Just listen to Hank Williams or Patsy Cline in the classic "Lovesick Blues."

Cows, in fact, may like music just as much as humans do. The internet is full of videos of people serenading cows. It is hard to tell if the cows are simply curious or whether they actually enjoy music for its own sake. What little research that has been done suggests that cows do like music. When music was played in a dairy, the cows were more relaxed and volunteered to be milked more readily, even though there was no effect on actual milk production.[21]

Of course I had to see how my cows would react to music. On a humid July morning of the third year, with acoustic guitar in hand, I hiked down to the lower pasture where the cows were grazing lazily. They eyed the foreign object in my hands with suspicion.

What to play? This was a momentous decision. After all, I was about to introduce the cows to music. The first song had to be just right. Ricky Bobby stood behind the others, as he always did when something new entered his zone. His sons, Walker and Texas Ranger, waited with anticipation. With their *Talladega Nights* provenance, the answer was obvious. I started playing the most recognizable riff in all of Southern rock.

"Sweet Home Alabama."

The reaction was, well, underwhelming. Nobody ran away, but only Daisy approached, and she seemed more interested in the guitar than the sounds that were coming out of it. Later that evening I tried a more systematic approach. Instead of playing guitar, I played music through a portable speaker. Sifting through different songs, I hoped to find something they liked. Not much seemed to interest them. Except, perhaps, anything with a horn section, such as "Soul Man" and "Tequila." Those perked up their ears. But I would not say that they found it enjoyable. The novelty piqued their interest, but only for a few moments, because then they wandered off.

I decided that those adorable internet videos were probably the result of conditioned responses, the cows having learned to associate music with something else they liked, probably food or treats or neck scratches from the farmer. If that were true, then it would be a simple matter to condition my cows to music. So I started playing Lynryrd Skynyrd's recording of "Sweet Home Alabama" every time they came up to the barn for grain and cattle cubes. The first time, they stopped dead in their tracks, and I had to coax them through the gate. After that, the music just became background noise. Critically, though, I always stopped the music when they were done eating so that music meant treats. Not the other way around.

It didn't take long. After a few days of training, they started coming over as soon as they heard the opening riff. The cows had become cultured, in a redneck kind of way.

CHAPTER 19

Cow Culture

Culture (noun): (a) the customary beliefs, social forms, and material traits of a racial, religious, or social group; (b) the integrated pattern of human knowledge, belief, and behavior that depends upon the capacity for learning and transmitting knowledge to succeeding generations.

–Merriam-Webster

W hen one thinks about culture, it is natural to call to mind the great accomplishments of human society: art, literature, music, and belief systems that transcend generations, such as the creation of democracy or religion. It is hard to imagine any animal, let alone bovines, having culture like we do. Indeed, culture is something considered uniquely human. But like most things that were once the sole province of humans, bits and pieces of evidence from far-flung corners of the animal kingdom have accumulated to the point that most scientists acknowledge that culture is intimately entwined with biological evolution.

If we let go of the human artifacts, like art and literature, culture boils down to knowledge and behaviors that are generationally transmitted. The other way information is passed from generation to generation is biologically—through reproduction and the transfer of genetic material. Historically, scientists have drawn a bright line between cultural and biological transmission, with only humans possessing the cultural pathway. The situation began to change in the 1950s, when instances of cultural transmission were discovered in animals.[1] The first, and most famous, was the spread of an innovative bird behavior in England. Titmice learned to pierce the foil top of milk bottles left on doorsteps and skim the cream. It began with a few birds but soon spread widely. This was not a hardwired behavior, so it had to be socially transmitted by birds observing one another.[2] In the 1960s, scientists discovered regional dialects of birdsong that depended on baby birds learning from their parents.[3] By the 2020s, evidence of cultural transmission had been found in a wide range of species, including humpback whales learning new songs, capuchin monkeys learning to use tools, vocal dialects in many bird species, fish that learned from their peers to avoid certain predators, and even bumblebees that learned to pull a string by watching other bees.[4]

To avoid the baggage associated with the term *culture*, scientists prefer to call these types of information transfers *social learning*. Different animals have different types of social learning. The simplest, and most common, is simply copying what another individual does. A more complex form of social learning occurs when an individual observes another's interaction with something or someone else. By watching if the outcome of the interaction is good or bad, the individual can learn by proxy. It is a very effective strategy because the observer doesn't incur any risk.

I had seen ample evidence of social learning in the cows' grazing behavior. Mostly they followed whoever was nearby, learning where the best forage was without going off on their own. On the rare occasions when researchers have tested the transfer of knowledge between cows (by watching cows that had been trained to push a panel to access treats or navigate around a barrier), no statistically significant effect was measured.[5] However, with a less demanding task that had negative consequences, cattle learned to stay away from an electric wire by observing other cattle avoiding it.[6] Researchers speculate that the different results in these experiments might be explained by the level of the cows' motivation. All animals are highly motivated to avoid pain, but cows may not be as motivated to obtain treats as, say, a dog. After all, cows have an unlimited supply of grass to graze. Thus, the level of social learning cows are capable of remains an open question.

Every herd is unique. Cattle are creatures of habit, and they get used to the daily routine of the farm. Imagine taking a calf from where he was born and relocating him to a new farm. Everything is different, including the other cows, who have established their own relationships and pecking order. Even the vocalizations sound different. It would be like moving to a foreign country.

Herd culture, then, derives from two sources: what humans do on the farm and what the cows do with one another. The two aren't completely independent because the cows react to the farmer and these reactions spread through the herd via social learning. On the flip side, the farmer also reacts to how the cows behave. So it is a two-way street. This is an important point, even if some farmers don't want to admit it. Cows sense how a person reacts to them. They might even provoke a person, just to test them. Every time Ricky Bobby head butted me, he was testing how I reacted. And not only that—all the other cows were watching too.

With this in mind, I became very conscious of the type of culture I wished to cultivate in the herd. Some of this would come from me, but I hoped to leverage the cows' natural ability to observe and learn from one another so as to foster a culture of kindness. If you gather up a group of animals—it doesn't matter the species—some will inevitably be nicer than others. Some will approach politely. Others will hang back. And then there will always be the one who acts like a jerk. Pushy. Jealous. Prone to sulking. To a large extent these are personality traits, but they also depend on the situation. Even a natural-born jerk can learn to temper his inclinations when social circumstances demand it.

Fortunately, I didn't have any assholes in the herd. They all had their moments from time to time. As the bull, Ricky Bobby had more than his share. Sometimes he would just get into a mood and stamp around, flinging his head side to side, turning over feeders or anything that wasn't anchored to the ground. The other cows would just stay out of the way. The mommas would gently nudge their calves to a safe distance until Ricky Bobby wore himself out. In this manner, the calves learned a bit of the herd culture: stay away from the bull when he was acting like a jerk.

The other thing the calves quickly learned was to nurse only from their own mommas. Nursing usually comes naturally, and after the initial latching on, each calf knows that it can count on its momma for milk. Most of this behavior is instinctual for both mother and calf. However, proper nursing is by no means assured. Every cattleman has stories of calves that wouldn't nurse. Sometimes it's because the calf is too weak or sick. Or the mother can't produce enough milk or, for reasons only the cows know, she rejects the calf and won't let it nurse. I was fortunate that none of those scenarios occurred. Calves being calves, though, they will inevitably think the proverbial grass is greener elsewhere and try to sneak a bit of milk from cows other than their mommas. Walker was particularly sneaky. He would sidle up to Cricket when she was nursing and try to latch on to a spare teat. Big mistake. Lucy brooked no intruders and would whip her head around and chase him off. But all the calves tried it at least once.

The calves were also schooled in the social order of the herd. Despite Ricky Bobby's bullheadedness, he was not really in charge. Lucy was the boss cow. She had been since the day she and Ethel and Ricky Bobby arrived, and probably even before that when they were with the old man. Lucy, with her stoic demeanor, was never the first to the feed trough. That was Ricky Bobby. Lucy simply walked up to what she wanted and, like Moses parting the Red Sea, everyone just moved aside. If they didn't, she would snort derisively and push them aside with her horns. It wasn't aggressive. It was just an emphatic reminder that Lucy was in charge.

Each year, the new batch of calves had to learn that Lucy was the boss. The first year, BB took the brunt of the schooling. Xena didn't suffer his fate because she was Lucy's calf and was protected by her mother's status, making her full name, Princess Xena, particularly apt. And if the other cows didn't mess with Lucy, you can be doubly sure they wouldn't touch her calf. BB, though, came from Ethel, who was decidedly lower in status than Lucy, which put him at the bottom of the herd hierarchy. His place was set on the first day, when Ricky Bobby and Lucy tag-teamed him, pushing him through the fence rails and out of the pasture. This happened a few times before BB learned to give those two a wide berth. Whereas Lucy was communicating her boss status, Ricky Bobby made clear that there could be only one bull, and that was him.

In the second year, Tex and Daisy had to go through the same acculturation process as BB and Xena had, but at least they had the benefit of observing their older siblings. I saw no evidence that Daisy had a special relationship with BB, her full brother, nor did I see anything special between Tex and Xena. They were all at least half-siblings anyway. With Tex's arrival, though, Xena's status fell a notch. Lucy had kicked her off from nursing when she was six months old, while BB continued to nurse until he was nine months old. Maybe the mommas had special affection for their bull calves, because Tex hung onto Lucy's teat until the same age. It was only because I forced him off through the application of the black pepper oil that Tex let go of Lucy's udder. Daisy suffered the same fate as BB, but her personality was such that she took it in stride. Nothing much seemed to faze her, and she had no problem staying out of the Lucy zone.

By the third year, Walker, Cricket, and Luna had an easier time navigating the herd relationships. Because there were three of them, they formed their own mini herd. They played together and slept together. Plus, the previous years' cohorts were more tolerant of the little ones than were the OGs. The little Nachos had had the fortune of being born into a great big extended family with all the permutations of siblings, half siblings, aunts and uncles, and even, in the case of Luna, her grandmother, Lucy. Maybe it was because everyone was related through Ricky Bobby's bloodline, but I didn't observe any rivalries between the Lucy and Ethel clans. No Jets or Sharks here.

As long as everyone observed these basic rules, they got along swimmingly. They grazed together. They slept together. They appeared to be happy and content. The only time they got snippy with one another was when I doled out cattle cubes. The bigger cows had a tendency to push aside the smaller ones, even their own calves. I discouraged this type of behavior by giving the treats to individual cows, and only when they exhibited good manners. No pushing. No head butting. And they should lift their head to receive a treat—a position from which they couldn't hurt anyone.

The demand for politeness was my contribution to the herd culture, but it was a two-way street. The cows also taught me how to behave.

Much of what they taught flowed from simply being a cow. They were resilient. They lived outside 24/7, in the heat, in the cold, rain or shine. And never did they complain about the weather. They could have. If they wanted shelter, all they had to do was stand in front of the barn and bellow. But they never did. They were living proof of the adage "There's no such thing as bad weather, just bad attitudes." I couldn't help but be affected by their placidity and easygoing approach to life. This wasn't an aspect of their culture; it was biology. Their digestive systems operate on a different timescale than that of a carnivore. Cows don't need to catch their meals. They just graze and chew and chew. Although a human can't subsist like that, I found that just sitting with them while they ruminated was more grounding than any form of meditation I had tried.

But to get to the point where I could sit with them, I had to be accepted into the herd, and that did require becoming schooled in the ways of the cow. This was a long process, for nothing the cows did was fast. They thrived on consistency. So the first thing I learned was to be consistent in everything I did. I came out in the morning at the same time, shoveled the manure, freshened the water, and gave some treats before leading them to the pasture. Repeat in the evening. Any deviation, anything new, was a potential threat to the cow, so changes to the routine had to be introduced gradually. This was how they learned to walk through the Bud box, by incorporating it into their daily rhythm.

As a result, I learned to slow down. Cows dislike sudden movements. To enter their personal space, I had to do so with a calm and relaxed demeanor. This required me to move at cow speed, slowing the pace at which I walked. If I had to do something with my hands, I did my best impression of a sloth, making only deliberately slow movements. As a corollary to moving slowly, I also learned that to enter a cow's personal space and to have physical contact with them, it was best not to approach them directly. If I crouched down and waited, they would come to me.

Going slowly does not come naturally to us humans. It certainly doesn't for me, and it was something that I continued to practice. But I had come to believe it was the foundation of cow culture and the most important skill they could teach us. Bud Williams had it figured out decades ago when he said, "Slow is fast."

My version went like this: *If you wait, they will come.*

CHAPTER 20

Pasture Ornaments

Ogres are like onions.

–Shrek (2001)

In a testament to the reproductive efficiency of cattle, I had gone from three cows to ten in two years. With Ricky Bobby castrated, the herd size was effectively frozen, but ten cows were still more than I had bargained for. I had long realized that I wouldn't be selling any of the cows. How could I? They were a family unit emotionally bonded to one another. Even momentary separations caused great distress. Needless to say, nobody would be sent to freezer camp. And as far as dairy went, only Xena would let me touch her udder. Her milk tasted sweet and would have made great cream, but she let me get only a squeeze or two at a time. To turn Xena into a dairy cow would require training her to stand still, and I would have to commit to twice-daily milking.

I kept returning to the day she was born and this adventure began. Crowley, the farrier who helped get Xena hydrated, was stuck in my head, asking the obvious: *What was I going to do with the cows?*

Three years in, I had converged on an answer, although I doubt it would have made much sense to Crowley. To get to this point, though, I had had to consider the alternatives. There weren't many.

With no utility as food, most farmers would consider the cows pets. This is not uncommon. Farmers often keep special animals as pets—maybe a bottle-calf they got attached to—with no other function than companionship. I suppose this could describe my herd, but designating them as pets diminished their existence, reducing the cows to mere objects that made me feel good.

Were they just pasture ornaments? That is what farmers call animals that no longer fulfill the utilitarian functions they were bred for. A common pasture ornament is an old horse, usually prone to lameness. People get attached to horses, and getting rid of them or, worse, euthanizing them becomes unthinkable. So they get put out to pasture, indefinitely. This is not necessarily a bad thing. If the owner takes care of them, pasture ornaments can have good lives. But without a purpose, pasture ornaments are also prone to neglect. They might get low-quality feed and minimal human contact. Old dairy cows, if they somehow avoid the auction lot, sometimes end up as pasture ornaments. One of my neighbors had three Guernsey cows perpetually in his front pasture. They got cheap hay in the winter and picked through weedy grass in the summer. They didn't appear malnourished, but I never saw humans interact with them. They just existed.

Maybe when I became old and decrepit and unable to physically interact with my cows, they would become pasture ornaments. But until then, I was determined to give some meaning to their lives. Before getting to that, it is important to consider the normal life of a cow.

In 2021 there were 101 million cattle and calves in the United States.[1] This number has been drifting down since the 1990s, but still, that's one cow for every three people. That's more than the number of dogs (about eighty million) or cats (sixty million). There is an important difference, though. Dogs and cats have relatively stable populations. With lifespans of about ten years, new animals comprise relatively small percentages. The cattle population, though, turns over every two to three years. Only 10 percent are milk cows. Another 5 to 10 percent are heifers destined to become milk cows. The rest are beef cattle in various stages of maturity. They go to slaughter when they are twelve to twenty-four months old. The age depends on what they eat, grass-fed cattle taking longer to mature than grain-fed. This means that thirty-five million head of cattle are slaughtered for beef each year and replaced by new calves.

The life cycle of beef cattle can take many different forms, but most originate in cow/calf operations. The general goal is to keep a stable population of cows that produce one calf every twelve months. Sex is a prime determinant of the possible outcomes for a calf. Since most farms can keep only one or two bulls for breeding, most of the male calves are castrated in their first week of life. These steers are usually left to nurse from their mommas until they are three to six months old. At that point, the farmer might separate them to wean the calf but continue raising him until he reaches a weight sufficient to bring a good price at auction. Or the calf might be sold as soon as he is weaned, to be fattened up elsewhere. These calves usually get sold to feedlots. Heifer calves go through a similar process, except some will be kept for future breeding.

In my three years in the cattle world, I had met many cattlemen, and each had a different approach to their cow/calf enterprises. Another neighbor had a small herd of ten and took his calves to the sale lot when they were about six months old. Since this was before they were sexually mature, he didn't bother messing with castration. His bull, a two-thousand-pound Black Angus, was friendly enough to pet as long as you were on the right side of the fence. Although he didn't have a name, I called him, ironically, Snowball.

After the seed-and-feed was under new management, they hired a woman who kept a herd of about ten. Whenever I went in for supplies, we swapped stories and pictures of our cows. She was in love with her bull, Cowboy, like I was with Ricky Bobby and BB. She did not sell her calves at auction, though. She and her husband slaughtered them when the animals were about one thousand pounds and sold "freezer meat" direct to consumer. The way this works is that someone buys either the whole animal or a share of the animal. Then the animal is sent to a local slaughterhouse for custom processing, which is exempt from USDA inspection, or processed on site. After processing, the owners pick up their steaks and hamburgers.

On a somewhat larger scale, Will Harris runs White Oak Pastures, a soup-to-nuts operation of one thousand cattle in south Georgia.[2] Grass-fed their entire lives, calves are raised naturally until they are two years old. They are slaughtered and processed on site. The meat is frozen and shipped direct to consumers all over the Unites States. This is not a trivial operation.

To sell processed beef across state lines, Harris had to build his own USDA-inspected processing plant—needless to say, a huge capital investment for what is still considered a small operation by industry standards. He employs over one hundred people—the largest employer in his rural county—and just making payroll is a constant challenge. The margins in the industry are razor thin. Having seen White Oak Pastures firsthand and spent some time with Harris, I wouldn't hesitate to buy his beef because I know the animals are treated well and live as they were meant to.

But these small-scale, local cattle ranches account for only 3 percent of the beef sold in the United States.[3] The rest comes from feedlots, aka concentrated animal feeding operations (CAFOs). If a CAFO has more than one thousand cow/calf pairs, then the EPA considers it large and subjects it to a number of regulations relating to the disposal of the huge amounts of waste.[4] The largest cattle CAFOs may hold over one hundred thousand animals. Despite the EPA's regulation of the pollutants these operations release, there is scant data on the number of animals being processed at these facilities, or even the number of CAFOs.[5]

Why is this important? Two reasons: environmental impact and animal welfare.

Cows get a bad rap for their effects on the environment. Worldwide, they are directly responsible for an estimated 10 to 20 percent of greenhouse emissions.[6] Most of this comes from nitrous oxide associated with the production of feed and, famously, the methane from cow burps and farts.[7] Both of these gases have worse effects on global warming than carbon dioxide. While the atmospheric effects depend on the total number of cattle in the world, CAFOs make the problem worse by the way the animals are housed and fed. They are called *feedlots* because the cattle are given prepared rations to put on weight and produce the highest-quality beef. The formulation of these rations is an exact science to put on weight as fast as possible. It is the cattle equivalent of living off of Hot Pockets, pizza, and Diet Coke. Whether it is for a cow or a human, a lot of energy goes into the making of processed foods. And what goes in determines what comes out. It is no wonder that feedlot rations, chock-full of nitrogen to build muscle mass and carbohydrates for energy, have the downstream effect of methane production. There are ways to mitigate some of the environmental effects, notably better grazing practices and improvements in feed digestibility, but they are not in

widespread use. Methane, for example, is a byproduct of carbohydrate digestion in the rumen. Eliminating grain and corn from the cattle diet would go
a long way toward decreasing their effects on global warming.

And then there is the animal welfare aspect of CAFOs. True, cattle are
herd animals. But there is a big difference between a herd of a hundred and
a herd of a hundred thousand. Cows, like people, form small networks. The
distress a young calf endures when he is taken from his momma and transported by truck or train to a feedlot in the Midwest, where he spends the
remainder of his days, is intense. If the CAFO is well-managed, his physical
health will be attended to. After all, nobody wants beef from sick cows. His
emotional life, though, will never be the same.

My intent is not to promote veganism. I occasionally eat meat, and the
food my dogs eat contributes as much to global warming as I do.[8] But there
is a difference in the morality of eating local beef versus that coming from
CAFOs. It has never been easier to buy meat directly from a farm. Yes, it
costs more, but none of the farmers I met were getting rich off of beef. Even
the fictional John Dutton of *Yellowstone* fame admitted that a good year was
one in which he broke even. The price of cattle at auction is determined by
their weight. Like all commodities, these prices fluctuate, and over the past
several years it has ranged from $1.50 to $2.50 per pound. Not all of the
cow is meat, but what farmers get is still a fraction of the price in the supermarket. Considering how long it takes to raise cattle and all the expenses a
farmer has, you quickly realize why cattle ranching is not a get-rich business.

While buying local meat might be the ethical thing to do, the approach
doesn't scale up to the amount of beef consumed in the world. There aren't
enough small meat-processing facilities to keep it local. And just because beef
is produced locally doesn't guarantee that an animal was treated well during
its lifetime. The largest US meat producers have begun to implement animal
welfare programs. Some advertise a commitment to the Five Freedoms: (1) free
from hunger and thirst; (2) free from discomfort; (3) free from pain; (4) free to
express normal behavior; and (5) free from fear.[9] However, there is currently
no independent monitor of animal welfare in the industry. Every time we buy
meat at the supermarket, we must acknowledge that CAFOs cannot ever allow
for normal behavior. And the act of slaughter, no matter where it occurs, cannot
ever be free from fear—the exception, perhaps, being the farmer who casually
walks up to his steer one day and puts a bullet in its head.

Short of giving up meat, consumers can still make educated choices of where they get their food. There aren't many meatpacking companies in the United States.[10] Contact them and ask them about their animal welfare programs, if any. Same for fast-food chains. At local restaurants, talk with the chefs. Tell them you support local farms who treat their animals well. The goal is to know where your food is coming from and how it is produced. Then you can make choices. Corporations pay attention to what their consumers want. All of this applies to dairy, too, but because dairy cows comprise a small portion of the cattle population, larger changes can be effected by focusing on the beef industry.

If Ricky Bobby and his crew weren't headed to freezer camp, what was to become of them? I couldn't shake the feeling that they were meant for something more than pets or pasture ornaments. Research subjects? Yes, that was part of it. All this time, I had been studying them like Jane Goodall had studied chimpanzees. These observations, in turn, shed light on what it's like to be a cow. Still, that was not enough to make life better for livestock or the people who worked on farms. Data points and observations alone weren't very good at changing people's behavior.

When people toured the farm, I always began with the story of why we left the city and how we fell into becoming cattle ranchers. Everything had a context. Understanding that these cows came to be not because I was raising them for meat but because they were helping me improve the land allowed people to see them in a different light. The cows were still being used as a sort of tool, but I considered it a symbiotic relationship. I took care of their physical and emotional needs, and by virtue of doing what came naturally—eating and pooping—they improved the soil. They also gave me a lot more than that.

They provided cow therapy. I interacted with the cows at least twice every day: in the morning and in the evening when I fed them. Whenever possible, I also visited with the herd in the middle of the day. After the morning graze, the cows settled down for their afternoon ruminations. They would lie down in a loose clump, chewing their cud, burping, and farting. This was when they were the most relaxed, and I often sat with them for a half hour or so.

A ruminating cow has a grounding effect—in both the literal and metaphorical senses. You become as rooted to the ground as they are. For the cow it is an act of pure digestion, and to be surrounded by it is a reminder that we, too, are biological reactors. In all my years of research, I hadn't found anything quite as relaxing as sitting with ruminating cows.

Ricky Bobby made for the best back prop. He offered the most mass to lean against, and he found contact with me as pleasant as I did with him. If he was deep into his ruminative session, I could sit down next to him and he would roll onto his side for a belly rub. There was a preferred side to this. Cows lie down in sphinx position with the forelimbs tucked under the brisket. But the position isn't symmetric. They will tend to roll slightly with the hindlegs splayed to one side. Because the rumen is on the left side, and rumination involves the continual burping and swallowing of fluid from the rumen, cows tend to tip to the right, which decompresses the rumen. That was the sweet spot to sit in—against the belly, just behind the left elbow. From there, I could lean back against his hulk and close my eyes or scratch the man-baby behind the ears. Ricky Bobby would often respond by tilting farther and farther to the right, until all four legs were jutting straight out into the air. Although Ricky Bobby rolled over the most frequently, all the cows except Ethel did it. I tried to spread the love among the cows during our afternoon sessions by spending some contact time with each of them. By the third season, BB was almost as big as Ricky Bobby and he enjoyed our cow therapy sessions just as much as the big man. Because BB had a gentler demeanor than Ricky Bobby, I could easily fall asleep leaning against BB without fear of being headbutted.

I often found myself drifting off into a meditative state. If someone had slapped an EEG on my head, I'm sure they would have seen me in an alpha rhythm. Alpha waves are eight to twelve hertz oscillations in the electrical activity of the brain and are observed when a person is in a relaxed state. Certain types of meditation may increase alpha waves, or, if a person enters a deep meditative state, the brain may slip into slower oscillations called theta waves.[11]

None of this was on my mind when I was sitting with the cows. (If it was, then I wouldn't have been in a meditative state.) After these sessions, though, I felt relaxed in a way I hadn't experienced before. Whenever I felt the weight of life's challenges, I sought out the cows. There was something undeniably therapeutic about sitting with several tons of cattle. The cows were well-tuned to picking up bodily states in one another, and humans too. They picked up on tension and moved away from it. Just entering their space demanded a certain level of self-awareness to be fully present and relaxed. The cows were gigantic biofeedback machines.

Ken felt the same way and came over for cow therapy once a week. After the evening feed, we would sit down together with the cows and relax with them until well after dark. Sometimes we just enjoyed the tranquility of a scene so pastoral it could make you cry. We'd marvel at the stars, a sight I had missed seeing from Atlanta. Venus and Jupiter did a *pas de deux* one year and seemed to always be in the western sky, while Mars crashed their party for a month or two before running off behind the sun. In the summer, we were serenaded by cicadas and the clicking of bats as they exited the barn loft. Barred owls called out plaintively, "Who cooks for you? Who cooks for you all?"

The natural beauty, though, was just the icing on the cake. We both came for the cows.

Ken told me how much the cows helped him. At first, this surprised me. Ken had been the head pastor of the largest church in town, and he was a big reason the congregation had grown to over ten thousand. He was loved immensely in the community—something I had seen firsthand as we traveled around the community, where he was still called Pastor Ken even though he had retired several years earlier. Before that, he had been a successful executive with U-Haul and served on the board of the Salvation Army Southern Territory Headquarters in Atlanta. Through his charity work, Ken was known by both politicians and business leaders throughout the state of Georgia. When Ken called, people answered.

Why, then, was Ken so entranced with the cows?

One evening, long after dark, he said, "You don't realize how much this helps me. Before you moved here, I had been feeling . . ." Ken's voiced trailed off as he searched for words. "Not giving up. Nothing like that. Cows have a special place in my heart. I grew up with them. I was even thinking about getting a steer to put out in our yard." Then ruefully he acknowledged, "But I'm just not physically up for it."

"Well," I said, "you know you're welcome here anytime you want. You don't have to ask me. Just come over."

"I appreciate that. These cows really uplift me. And I know that sounds crazy."

It didn't. Three years earlier, it might have struck me as odd. But no longer. "They lift me up too," I replied.

What was crazy, and for which I had no explanation, was the sequence of events that brought our life paths in juxtaposition. Before moving to the farm, I could not have imagined someone more different from me than Ken. He was fifteen years older, which, although not quite a generation apart, felt like it because of our vastly different life experiences.

Ken had done two tours in Vietnam and paid heavily for it in the form of PTSD and a raft of neurologic disorders thanks to Agent Orange. Me— well, let's just say I came from a long line of men who had somehow avoided military service. I had had the fortune to come of age during a particularly peaceful period in American history after the draft had been abolished. Ken grew up working on farms in rural Georgia and North Carolina. His parents' formal education ended in the third grade. I spent my youth on the beaches of Southern California with two college professors for parents. Ken was a well-loved pastor at the largest Baptist church in the county. I had followed in my parents' footsteps and was a professor at a top-rated private university in Atlanta. Religion had not been part of my life.

On the surface, my relationship with Ken could be the start of a bad joke: "A pastor and a scientist walk into a bar . . ." After three years, though, I considered Ken my best friend. I think he felt the same way because he once told me that you could count the number of true friends in your life on the fingers of one hand.

Maybe we would have become friends without the cows. But not in the same way, I think.

There are dog people and there are cat people. It's the same with farm life. There are horse people, chicken people, goat people—and then there are cow people, which is one of the few groups that has its own word: *cattle-man* or maybe *cowboy*, but that's a whole different thing. While you could be a farmer with some cows, devotion to their care and preoccupation with maintaining their pastures is what makes a cattleman. And a lady cattleman, who is a *cattlewoman*, is an even higher calling. The Georgia Cattlewomen's Association said it best:

We love our cows and tend to them like they are our children. There is nothing like seeing a calf born, watching it stand up for the first time, and go to its mother's udder for milk. It is Nature at its best. . . . It takes quite a woman to get out there and work cows, hay, and whatever else needs to be done on the farm. After all this is done, we come in the house and take care of our families. All these qualities and traits plus way much more is what makes us a Cattlewoman.[12]

Being a cattleman or cattlewoman is more than a job. It is a commitment to a way of life. It requires more than simply feeding the cows. Cattle live off the land, and to keep them happy and healthy a person must be willing to nurture the land. Although I had originally acquired the cows to help make life easier in managing the pastures, in every conceivable way, my life was more complex because of them. Whereas I could have been content to mow the pastures, I now walked them every day contemplating the management decisions that would sustain the herd and keep the cows healthy. And there was no avoiding the dirty, sweaty, physical work of shoveling manure, hauling hay, and repairing fences.

Ken was a cattleman, and now I was one. It was why we had become such fast friends, and it was the reason the cows provided succor to both of us. Could the cows help other people too?

Animals have been part of the fabric of human life for millennia. It was only when people moved to cities that they became disconnected from their animals. But because of our coevolution with animals, especially domesticated animals, they are intertwined in our DNA. Humans were responsible for the selection of traits that, for a variety of reasons, we found useful, known as the three Fs: food, fiber, and friends.

When it comes to animal friends, most people think of cats and dogs. But why not cows? Well, because they are eaten, and nobody wants to think about eating a friend.[13] Most dog and cat owners consider their pets as family members—that is, family members who never say mean things or who want to borrow money or nag you to take out the garbage. And while it may be a stretch to say that our pets love us unconditionally, there is plenty of evidence that they do, in fact, have emotions similar enough to love to use the L-word.

Whatever it's called, the healing power of the human-animal bond has

been recognized for thousands of years. Florence Nightingale famously brought animals into the modern era of medicine when she treated casualties of the Crimean War, writing, "A small pet animal is often an excellent companion for the sick, for long chronic cases especially."[14] Ever since, animals have been used in a growing range of therapeutic roles. Beyond therapy, some animals can be trained to provide assistance to people with mental and physical disabilities, in which case they are called *service animals*. The vast majority of service animals are dogs, but both dogs and cats act as *therapy animals* and *emotional support animals*. Therapy animals are usually brought to people in institutional settings like hospitals and nursing homes, where they provide comfort to patients. In contrast, emotional support animals live with an individual. Dogs are the most common therapy animal but certainly not the only one. Horses are employed for equine therapy, and a variety of barnyard animals has also been used, including rabbits, pigs, donkeys, and alpacas. So why not cows?

Although I had yet to see a cow wearing a vest emblazoned with Therapy Cow, there were, in fact, a few niche farms offering just that. The Gentle Barn, with locations in California, Tennessee, and Missouri, offered cow hugs. In the Netherlands, they called it *koe knufflen*, for "cow hugging," which the BBC had dubbed the latest wellness trend, along with goat yoga.[15]

All of these human-animal interactions fell under the umbrella of animal-assisted interventions (AAI). Although the therapeutic effects of AAI had been recognized by some in modern medicine, many clinicians continued to dismiss it as a placebo effect. After all, it was difficult to design placebo-controlled trials for efficacy. The situation began to change in the early 2000s, when researchers started looking more closely at objective measures of wellness, like heart rate, blood pressure, and levels of stress hormones. There was also a shift from using animals to treat physical disorders to mental ones like autism, depression, and PTSD.[16] Even so, AAI was still a niche area with little money available for research.

One of the themes in AAI that emerged from this research was that animals play a uniquely nonjudgmental role in therapy. Animals don't judge people by the way they look, how they sound, or where they come from. As one researcher put it, they act as a secure base to the here and now.[17] When people are more mindfully present, it opens a door to other experiences, like connecting with other people.

But for any of these purported benefits to occur, there had to be some methodology. You can't just stick a person with a dog—or a cow—and expect a magical healing to occur. Not all animals are suitable for this kind of work. And even with the right animal, the person has to be instructed in what to do. Some people might benefit from quiet physical contact, while others need something more interactive. Most of the research has ignored these sorts of individual differences in both the animals and the humans.

With this in mind, I reached out to Ken to help create a therapeutic program around the cows. He had no shortage of friends who were in need of peace and tranquility. And I would put what I had learned about the cows to good use in preparing both the cows and the people for these sessions.

It was important to brief participants. Everyone has preconceived notions of how cows behave, so I had to disavow them of these beliefs and open their hearts and minds to the possibility that cows have the capacity to love humans, as dogs do. The briefing was designed to take no more than ten minutes, and it hit the high points of the intelligence of cows; the importance of moving slowly and making a good first impression (cattle cubes help); where to touch them; how they show affection; and how to avoid getting injured.

The first intrepid visitor lived down the road. Ken was vague about why he had invited him other than to say, "Jerome could definitely use some relaxation." It made no difference to me the reasons a person might benefit from sitting with the cows. I just hoped the cows would cooperate with a stranger.

Initially, the herd hung back. They didn't run away, but they didn't approach either. A bucket of cattle cubes quickly changed their minds, and Jerome was feeding them as fast as they would take them. Before long, the cows began to settle down for their afternoon rumination. As long as I was there, they felt safe, and so Ken, Jerome, and I sat down with them. It was quite remarkable really.

The fact that the herd let us do that, and the impression it made on Jerome, was everything I had hoped for. We stayed there for half an hour, three guys in a pasture with a herd of cows, not doing much of anything.

To say that we didn't do anything doesn't mean that nothing happened. To my eye, something magical occurred.

When you sit with the cows you notice things that you wouldn't normally pay attention to. You are surrounded by the chewing of grass. The cows radiate warmth from the fermentation in their rumens. The cows connect a person to the environment in ways that no other animal does. This, I think, is a big part of their therapeutic effect. Dogs and cats don't do this. Their effects on humans come directly from them, such as the purring of a cat or the soft snoring of a dog in a person's lap. Even horses have a different relationship with humans than cows do. A horse's therapeutic effect comes from the very intimate relationship of a person sitting astride its back. Horses rarely lie down, so it would be a rare occasion for a person to sit down with them. Cows, though, lie down so much they are almost inviting you to join them.

Several months later I ran into Jerome at the county dump station. Like Ken, he had grown up around cows but had never seen fit to sit down with them, so our pasture session was the first time for Jerome. I asked him what he had thought about the experience.

"Peaceful," he replied. Jerome was not a talkative sort, and I didn't expect any more than that, but after a few beats he smiled and asked, "How's Luna?"

"Growin' like a weed!"

It was a completely ordinary exchange between two guys in the country. Three years earlier I would have dismissed it as meaningless small talk. But like Shrek, we were all onions, and each interaction peeled back a layer until eventually we got to our cores. Not many people knew who I was. Sure, many folks around the county had probably heard about the weird professor raising mini cows, but until they actually came out to the farm and saw how the cows were helping improve the land and that they were gentle, lovable creatures, too, they didn't really know me. By opening up the farm and the cows to other people, I shed a layer or two of my onion skin. The cows helped me become part of a community. It was something I had never experienced.

Ken had always been about the creation of community. The cows were the catalyst, just like they had been for our friendship. They had a demonstrably calming effect on everyone who visited the farm. If you slowed down to the cows' pace and met them on their level, they welcomed you into their world and connected everyone who took the time.

People of all ages came. A colleague at the university came to the farm with his wife and fifteen-year-old daughter, who sat with Xena for a good fifteen minutes. A young couple, newly transplanted from California, moved in down the road. We became good friends over our shared interest in sustainable agriculture, and they got quite attached to Cricket. Whenever they had out-of-town visitors, they brought them over to meet the cows.

One of Ken's older friends, an eighty-eight-year-old widower, who had also grown up with cows, came to the farm to see for himself this wonder of which Ken spoke so highly. When Luna stuck her head through the fence rails for a neck rub, he exclaimed, "I never seen anything like this! What kind of cows are these?"

I laughed and said, "Cowpuppies."

APPENDIX A

A Brief History of the Cow

A million years ago, long after the dinosaurs had gone extinct but before humans appeared, the great mammals ruled the earth. This period, the Pleistocene, saw the rise of some of the most spectacular animals that ever walked the planet: saber-toothed tigers, dire wolves, giant ground sloths, marsupial lions, woolly mammoths—and the ancestors of cattle: the aurochs.

The aurochs was an imposing beast. An adult bull stood six feet at the shoulders and weighed over three thousand pounds.[1] Three-foot-long horns arched forward from a massive head that could ably fend off wolves and bears. During mating season, the bulls used their horns to fight each other, sometimes to the death, for the right to breed with the females.

Early humans had great regard for the aurochs too. The Lascaux cave paintings in southwestern France depicted numerous aurochs as humans seventeen thousand years ago saw them. The most famous of these compositions was found in the Hall of the Bulls. Rows of aurochs faced each other. Some were accompanied by horses. One of the bulls was over seventeen feet long, the largest of all the images in the caves.[2]

A few thousand years after the cave paintings were made, the earth's climate changed. The glacial ice sheets retreated north, and by twelve thousand years ago, all of the great mammals had disappeared. But not the aurochs. They hung on until 1627, when the last one died in a reserve in Poland.[3] Because they lasted so long, this gave rise to the unique situation of the wild ancestors of cattle coexisting with their domesticated counterparts for thousands of years. The final aurochs, though, were much tamer than their ancestors.

Biologists classify quadrupedal mammals primarily on the number of toes. The order Artiodactyla includes all the even-toed (aka cloven-hoofed) ungulates. The division of animals into even- or odd-toed families is based on the number of weight-bearing toes. In actuality, all mammals have five toes, but in many species some of the toes are mere vestiges, the so-called dewclaws. The artiodactyls include the aurochs and its cattle descendants, plus a wide range of other species like giraffes, hippopotamuses, camels, deer, sheep, and goats. The odd-toed ungulates, the perissodactyls, have either one hoof (horses, donkeys, and zebras) or three (rhinoceroses and tapirs). Apart from the toes, the two orders also differ in their digestive systems. The artiodactyls have multichambered stomachs and are considered true ruminants, whereas the perissodactyls have only one large stomach that evolved to digest plant material without the need for regurgitation.

Further down the tree of mammals, the aurochs belonged to the Bovidae family, which includes cattle, sheep, goats, gazelles, antelopes, and wildebeests. They share basic traits like a blunt snout, a muscular neck, and horns on the males. Horns are permanently affixed to the skull whereas antlers, as on deer, are shed each season. In some species, like the aurochs, the females also have horns.

Before their decline, the aurochs roamed Europe, Asia, and North Africa. The North African aurochs was not domesticated and eventually died out, but the Asian aurochs was domesticated in India around nine thousand years ago and the European aurochs six thousand years ago. The Asian and European lines continued on mostly separate evolutionary paths, giving rise to the two modern species of cattle: *Bos indicus* and *Bos taurus*, respectively. Because the Asian cattle—commonly called *zebus* or *Brahmans*—evolved in the hot, humid climate of India, they became well-adapted for higher temperatures and were valued for their hardiness during droughts. They are easily recognized by the prominent hump between the shoulders and large dewlaps that function as radiators, like an elephant's ears. The European cattle—taurines—became adapted to cooler temperatures and harsher winters by developing longer coats. They also had more muscle mass because large animals are more metabolically efficient in retaining heat (but have more difficulty in cooling themselves in hot weather).

How the aurochs was first tamed, and then domesticated, remains a mystery. The period of domestication, when humans transitioned from being hunter-gatherers to living in semipermanent communities, growing crops and animals, predated the invention of writing. So there is no Rosetta Stone to indicate how it was done. The most likely scenario is that when humans figured out they could cultivate crops, especially cereal grains, they discovered that grazing animals were also attracted to them. The tamer aurochs would have hung around the fields of wheat and rye (not corn, which was first cultivated in North America), grazing the leaves when the humans weren't around. It wouldn't have taken long before the people realized that they could let the ruminants graze extra plants and then slaughter them for meat, which was far easier than hunting them. Naturally, only the tamer and least aggressive of the aurochs would have been allowed. Scientists and archaeologists still don't understand how this process led to the change in body morphology that defines modern cattle. Bones and carvings, however, indicate that by six thousand years ago, zebus were widespread throughout Asia and taurines in Mesopotamia. When people migrated out of these birthplaces of human civilization, they took their animals with them. Dogs, of course, but also cattle, sheep, and goats.

As humans moved about the globe, subgroups of cattle evolved, turning into the different breeds. There may be as many as one thousand breeds of cattle, although it is hard to know how many are still in existence. Each breed reflects its origin—zebu or taurine—and its function.

There are three main uses for cattle: beef, dairy, and work. Beef cattle are muscular and reach maturity quickly because the faster they reach slaughter weight, the fewer resources they consume. Dairy cattle have bigger udders, with some, like the Holstein, producing immense quantities of milk. Working cattle, or oxen, are trained to pull things like plows and transports. They have become less common as these jobs have been handed off to horses and, later, motorization.

In the course of domestication, a remarkable change occurred: some of the cattle lost their horns. This provided an obvious improvement in safety. Cattle that never develop horns are said to be *polled*. Cattle with horns, which are later removed, are said to be *dehorned*. Some of the earliest references to polled cattle are found in carvings on Egyptian tombs from 2000 BC. The most famous, the Weeping Cow of Kawit, depicts a young man milking a cow that is shedding a tear. A calf is tied to the mother's leg. The weeping cow has no horns. Maybe they were cut off, or maybe she was born without them. Archeological evidence, which can tell the difference, points to the development of hornless cattle well before the tomb of Kawit, perhaps 3000 to 4000 BC, suggesting that the Weeping Cow was, in fact, polled. During the Middle Ages, polledness spread through the taurine lines. Scottish breeds, like the Angus, became well-known for their lack of horns by AD 1000.[4]

The development, or nondevelopment, of horns was itself an important aspect of what it means to be a cow. Calves are not born with horns. The horns begin developing during the first few months of life, originating in a nub of skin tissue on top of the head. These horn-producing cells make keratin, which grows both upward and inward. By the time the cow is a year old, the horn tissue begins to attach to the skull itself. Over the next year, the base becomes contiguous with the skull and contains a large, blood-filled sinus. Somewhere in the course of domestication, mutations occurred that resulted in the absence of the horn-producing bud cells. Genetic analysis has homed in on a region of the first bovine chromosome as causing polledness, which is dominant over the natural form. So if one parent has the polled gene, at least half of the offspring will too. The polled gene undoubtedly causes effects outside of horn development. Some polled breeds are prone to developing double eyelashes or the dreaded corkscrew penis.[5] The extent to which the mutation affects cattle remains an area of ongoing debate.

Modern agriculture favors a homogeneous, commoditized product. Because horns were dangerous in an industrialized setting, polled cattle soon became dominant. In the US, a dozen or so breeds account for most of the beef population. These include the ubiquitous Black Angus, Hereford, Charolais, and Simmental cattle. The dairy industry favored milk quantity above all else, with the Holstein, Jersey, Guernsey, and Brown Swiss being the most common breeds. As a result, industrialization of agriculture has led to the decline of many other breeds. Some no longer exist. Others have become so rare that they are designated heritage breeds, which is a sort of endangered species status.[6] The modern cow, then, is no longer the same as pre-industrial cows.

The same can be said of all domesticated animals. Dogs, horses, goats, and the rest of the farmyard menagerie have all evolved along with humans' ever-changing needs, but dogs are truly the jack-of-all-trades. Their rapid reproduction cycle allowed humans to breed them into all sorts of shapes and sizes to fit every niche that people could imagine. Above all else, dogs have to get along with humans and, to the extent possible, understand language. Cattle also have to get along with people, but only to the extent necessary for ease of handling. Cattle ticked only two of the three Fs: food and fiber, but not friends.

And yet people still become attached to cows. In Hinduism, the cow represents divine and natural beneficence. They are considered a gift from the gods because they provide milk and butter and cheese without asking for anything in return. Dairy cows often have special relationships with their caretakers. It is easy to understand why. The farmer has to milk them twice a day. Any animal requiring that much attention had better have a nice temperament. It is common for a family's milking cow to have a name. Beef cows, not so much—there's too much cognitive dissonance when we eat named animals.

Our cows were zebus, indicating that they were descended from the Asian lines. Pure zebus weren't very common in rural Georgia, which was strange considering the climate. Zebus were well adapted to the long hot summers and would thrive there, but taurine cattle dominated the landscape. Black Angus were everywhere, as were their English cousins, the Herefords. The only trace of zebu in these parts was a cross between zebu and Angus, called a *Brangus* (Brahman + Angus). Most of these cattle are sold at local auctions and shipped west to feedlots in Kansas and Nebraska, where they are fattened up before slaughter. Buyers prefer specific breeds because that is what they know. The ranchers just supply what the market wants, even if other breeds were better adapted to the local climate.

By virtue of their humps, our cows appeared exotic, but their diminutive size made them downright weird to my ranching neighbors. It shouldn't have. All the domesticated animals have miniature versions. Dogs, of course, run the gamut, from the Great Dane down to the teacup Chihuahua. The horse has yielded small breeds like the Shetland pony. In the swine world, there is the pot-bellied pig. With cattle, there is the Dexter, a smallish Irish breed favored for being an easy keeper on small acreage; the miniature zebu; and its taurine cousin, the miniature Scottish Highland, sometimes referred to as the Ewok of cows because of its fluffy coat. Making tiny versions of these animals wasn't difficult—just keep breeding the smallest animals to each other until you get down to a manageable size. Regardless of the species, minis served one purpose: companionship.

Because cattle have been part of human society since the dawn of civilization, they have exerted strong influences on cultures all over the world. Cultural references generally reflect the duality of cattle as being both fearsome, as embodied by the bull, and maternal, as in a cow that provides sustenance. The Epic of Gilgamesh, carved into stone tablets four thousand years ago and considered the oldest written story, tells the tale of King Gilgamesh slaying the Bull of Heaven. The goddess, Ishtar, had fallen in love with Gilgamesh, but after he refused her advances, she sent down the bull to kill him. Instead, Gilgamesh slayed the bull and threw its leg back toward the heavens, where it became the constellation Taurus. The agricultural importance of the heavenly bull coincides with the onset of spring as the sun moves into the constellation.

Similar stories could be found in other cultures. In Greek mythology, Poseidon sent a snow-white bull to King Minos of Crete. The bull was too beautiful to fulfill the sacrifice that Poseidon had been expecting. Without the sacrifice, the angry god took revenge by causing Minos's wife to fall in love with the bull. She later gave birth to a half-man, half-bull, called the Minotaur, which was so fearsome that a labyrinth had to be constructed to hold it. For his seventh labor, Heracles took the white bull away, but it broke free and wreaked havoc throughout Athens, causing war with Minos. Theseus eventually captured the bull and went to Crete to slay the Minotaur.

In Mesopotamia, the *lamassu* was a mythical creature with a human head atop a bull's body with the wings of an eagle. Large sculptures can still be found throughout the Middle East.[7] In Hinduism, Shiva uses a white bull for transport, called Nandi, which is also the Sanskrit word for happiness. In Buddhism, protecting cattle (and other animals) is considered good karma because they might be reincarnated human beings. Even in modern times, the bull continues to symbolize strength, fertility, and stubborn determination. The Bull of Wall Street, for example, embodies all of these traits.

In contrast to the bull, female cows embody motherhood because they provide milk. Hindus speak of Kamadhenu, the mother of all cows. She takes the form of a white cow with a woman's head and lives in Goloka—the realm of cows—with the god Krishna. Kamadhenu produces milk for Krishna as well as warriors to protect him. Ancient Egyptians worshipped Hathor, the goddess of women and love, who also took the form of a cow. In the *Edda*, the Icelandic equivalent of the Bible, primordial beings existed in a sort of void before Odin created humans. Ymir was one of these primeval beings who condensed into existence from poisonous water dripping out of a frozen river. For sustenance, Ymir suckled at the teats of the primeval cow Audumbla.

In Greek mythology, stories of cows were not as common as those of bulls, but Hera, the goddess of marriage and protector of women, was sometimes called *Boōpis*, meaning "cow-eyed," perhaps because of her jealous nature. In the most well-known story, Zeus became infatuated with a beautiful mortal woman named Io. Accounts differ as to who was responsible, but Io ended up turned into a heifer. Some said Zeus did it to hide her from Hera. Others said Hera herself did it. Io fled to Egypt, where she eventually regained human form and bore several of Zeus's children.

In modern times, the reverence for both bulls and cows is reflected in art. Seventeenth-century Dutch painters were particularly taken with cows, reflecting the importance of cattle and Dutch pride in milk production. Some of the most famous painters of the eighteenth and nineteenth centuries featured cattle in their works too. Van Gogh, known for his portraits and landscapes, did several cow paintings, which was unusual because he almost never painted animals, and even then, they were rarely the focus of the painting. In fact, the cow was the only animal he painted as his subject, appearing in five works.[8] In the twentieth century, Picasso featured cattle, especially bulls, in both paintings and sculptures. A bull was part of *Guernica*, and he even did a series of lithographs, *Le Taureau*, aimed at capturing the essence of a bull with as few lines as possible.

The American relationship to cattle is somewhat different than the European one. Although cattle serve the same purposes in the United States, mostly beef and dairy, it was the conquering of the West in the 1800s that so defined American's image of cattle, epitomized by the cowboy.

When the Spanish began settling the Americas in the 1500s, they brought cattle and livestock. *Vaqueros*—from the Spanish word for cow, *vaca*—were hired by the early ranchers to manage the herds. Many of the vaqueros were Native Americans and Mexicans. From Mexico, the ranches spread south into Central and South America and north into what would later become Texas. When the Civil War broke out, Texas joined the Confederacy, leaving the herds largely unmanaged as the ranchers headed east. Those who survived returned to find their herds multiplied severalfold. The North, depleted of resources, was hungry for beef. A livestock trader from Chicago, Joseph McCoy (the "real McCoy"), realized that a new railroad in Abilene, Kansas, could transport the surplus beef east.[9]

Cowboys drove the herds from Texas to the railheads in Kansas along the Chisholm Trail. All over the West, cow towns sprung up along other trails: Wichita, Fort Worth, Dodge City, Cheyenne. Cowboys weren't paid until the end of a successful drive, so they went on spending sprees in these towns, resulting in their predictable reputations for drinking, whoring, and fighting. With the invention of barbed wire in the 1880s, the open ranges began getting chopped up into grids. Homesteaders and wealthy ranchers claimed previously public lands, and the great cattle drives came to an end. Cowboys still exist, but now they live and work on private ranches instead of the open range.

Rodeo remains as one of the artifacts of cowboy culture and has increased in popularity in recent years. The professional rodeo circuit has calf roping, steer wrestling, and barrel racing, but the pinnacle of rodeo will always be bull riding. With their big hats and belt buckles, the riders get all the glory. None of this would be possible without bulls capable of tossing a rider like a horsefly on the rump. These aren't just cantankerous bulls; they are athletes bred and trained for one thing. The owners of the best bulls win prize money just like the riders. The time and attention required to create a world-champion bull also fosters close bonds between the trainers and the animals. Of course, these bulls have names, like Cool Whip and Ridin' Solo, and they have the privilege of living out natural lives on their ranches.[10]

Most people, though, no longer have the opportunity to interact with livestock. How can they? In 2020, 80 percent of the US population lived in urban areas.[11] This isn't a new phenomenon. Every decade since the census began in 1790, the percentage of people living in or near cities has increased. We long ago lost touch with where our food comes from and the fact that the animals who provide for us are individuals with personalities of their own. Even in the canon of Western literature and movies, cattle are treated as interchangeable objects.

Cow Experience Briefing

How cows think

- They are intelligent.
- They recognize individuals and identify them as friend or not. So make a good first impression.
- As prey animals, they are afraid of anything unfamiliar.
- They are highly social and emotional.
- They show affection by licking one another. This is usually on the head and neck, so don't be surprised if you are licked. Their tongues are strong and very rough and are used like an elephant uses its trunk.

Movement and noise sensitivity

- Move slowly, like a sloth.
- No loud noises or squealing. Talk in low, soft tones.

Body language

- Head up is relaxed and the position we prefer for interaction.
- Head down can mean different things. If they are relaxed, they might just want the top of their head scratched. If the chin is tucked in, then they are testing whether you can be moved and you are about to be head-butted. Do not allow this.

How to approach

- Cows use their head and horns for protection.
- Don't approach from head-on or from directly behind.

- Approach at an angle or from the side.
- If they start to back up, stop and crouch down.
- If you crouch down, they will eventually come and investigate.

How to get away

- Don't ever turn your back on a cow!
- Exit at a right angle so you can keep one eye on them.
- If they start to head-butt you, do not back away. They will just keep pushing you. Instead, take a decisive step toward them. If they don't back off, then walk quickly past their flank until you are past the rear. If they come at you again, move to the side and let them pass (like a bullfighter).

Glossary

Abattoir. Facility where animals are butchered. From the French *abbatre*, meaning to strike down.

AI. For cattlemen, artificial insemination. But in the new era of farming, artificial intelligence.

Animal Unit Equivalent. One AUE equals a single one-thousand-pound cow. AUEs are used to determine how many animals of different sizes and species can be managed on a given parcel of land. See *carrying capacity.*

Bagging up. When a cow, or heifer, is close to calving, the udder swells with the first milk (see *colostrum*). Also called *udder drop.*

Bottle-calf. A calf that has been bottle-fed. Dairy calves are pulled from their mothers and bottle-fed, and occasionally beef calves are rejected by their mommas and need to be bottle-fed. These calves become highly socialized to humans.

Bud box. Named after the most famous stockman, Bud Williams, a rectangular cattle pen with a ninety-degree alley at the midpoint of one side. What makes it a Bud box is how the handler works the cattle inside the box, flowing first to the far end. The handler takes advantage of cows' instinct to head back where they came from and diverts them into the chute.

Bunting. Rubbing or butting the head against an object or another animal. Can be a form of play, aggression, or scent marking. Calves will bunt the mother's udder to induce milk flow.

Bush Hog. An implement for mowing. Stout blades are powerful enough to take down bushes. Also called a *brush hog.*

Calf slippers. The waxy substance covering the hooves of a newborn calf, called *eponychium*. It protects the momma from injury during birth but hardens within minutes of being exposed to air.

Carrying capacity. How many animals a plot of land can support. Often expressed in AUEs per acre.

Cattle cubes. Morsels of compressed alfalfa used to supplement cows that are grazing natural forage. The cubes are actually cylinders about a half-inch in diameter and one inch long. Also called *range cubes*.

Colostrum. The first milk, which is loaded with antibodies. It is critical that the calf get plenty of colostrum within the first twenty-four hours, preferably in the first six hours. After one day, the calf's digestive system is unable to absorb the antibodies.

Coping style. How an animal reacts to unpleasant stimuli. Examples: fight or flight; approach or avoidance; boldness or shyness.

Cud. The wad of grass a cow burps up from her rumen and sticks between her cheek and gum for further chewing.

Cultipacker. Implement that tamps down soil after seeding.

Dewlap. The flap of skin hanging down from the neck.

Disbudding. The process of removing the horn buds of a newborn calf that is not polled. Two methods are commonly employed: high temperature cauterization and chemical abrasives.

Emotional support animal. An animal that provides emotional support, alleviating one or more symptoms of a person's disability. They provide companionship, relieve loneliness, and may help with depression and anxiety but do not have special training to perform tasks. See also *service animal*.

Fly rub. An insecticide-impregnated sock suspended between two posts. Cattle walk beneath it, rubbing their backs with insecticide that kills flies.

Five Freedoms. Internationally accepted standards of humane animal care: (1) freedom from hunger and thirst; (2) freedom from discomfort; (3) freedom from pain, injury, or disease; (4) freedom to express normal behavior; (5) freedom from fear and distress.

Food plot. An area, often carved out of forest, for growing plants that wildlife like to eat. Although conservation is sometimes the goal, a food plot is used for attracting and fattening up game wildlife, especially deer and turkey, which will be harvested during hunting season.

Freezer camp. Euphemism for the slaughterhouse. See *processing* or *abattoir.*

Grafting. The process of getting an orphaned calf to nurse from a cow other than its mother. If the surrogate has lost her calf, then she may be fooled into accepting the orphan by smearing it with birth fluids.

Grazier. Someone who manages pastures for livestock.

Harrow. Implement for turning up topsoil. A series of discs is pulled behind a tractor. The angle of the discs can be adjusted from straight on, which results in a light soil disturbance, to an aggressive angle of about twenty degrees, which acts almost like a plow.

Hay. Grass that has been cut and dried for storage. The oldest way to store hay is to rake it into a pile, called a *shock*. Baling allows hay to be transported from the field and stored elsewhere, ideally under cover. Bales can be either square or round.

Horn flies. Small biting flies that cluster around the head, neck, and shoulders. Severe infestations cause cattle to lose weight because they spend so much time trying to swat them away instead of grazing.

Hotwire. A strand of electrified fencing. Two strands are enough to keep cattle contained, although they should not be the sole source of fencing because if they get broken, the whole fence becomes unelectrified.

Implement. Any tool attached to the rear of a tractor.

Input. Anything brought onto a farm to grow things. Common inputs include fertilizer, feed, and fuel. From an economic standpoint, the value of the agricultural produce (the outputs) should be greater than the cost of the inputs for a farm to make a profit.

Low. Cow vocalization. Synonymous with moo, although cows don't bring their lips together to make the /m/ sound of *moo*.

Management-intensive grazing. An approach to managing grazing animals that depends on frequent movement. Subdividing pastures into smaller paddocks and moving the animals on a daily basis allows for brief, intense grazing followed by long periods for the forage to regenerate.

Muck. A slurry of urine, feces, and whatever they land upon. Urease, an enzyme produced by the bacteria in manure, reacts with the urea in the urine, converting it to ammonia. Cleaning up the mess is called *mucking out* a stall, for which it is prudent to wear *muck boots*.

Neoteny. The retention of juvenile traits or behaviors in the adult of a species. The outstretched neck of a cow receiving licks is a neotenous behavior mimicking the posture of a nursing calf.

Nitrogen fixer. A plant that can take atmospheric nitrogen and convert it into a form that becomes available for growth in the soil. Legumes, like soybeans and clover, are common nitrogen fixers used in agriculture.

No-till drill. Implement for planting with minimal soil disturbance. The first row contains a series of discs (like a harrow at zero angle) to slice through dead grass and mulch. The second row has pairs of discs angled in a *V* to open up the slit created by the first row. Tubes connected to a seed box drop seeds into the *V* at a metered rate. The third row consists of rubber discs held under pressure by springs. These close up the soil, ensuring good seed-to-soil contact.

Open cow. A cow that should have been impregnated but wasn't.

Personality. An individual's behavioral and emotional traits that are consistent over time and different situations.

Polled. Cattle that do not have horns. A single gene controls the development of horns, with the poll allele dominant over the recessive horn allele.

Processing. Euphemism for slaughter.

PTO. Power take-off. A splined hook-up on the rear of a tractor for attaching the drive shaft of an implement. Most implements are designed to run with a PTO speed of 540 RPM. The PTO and the accompanying drive shaft are notorious for catching clothing and hair, with predictably disastrous results. Modern tractors have safety features that automatically shut off the PTO if the driver's seat is empty. Implements should have a nonspinning shield around the driveline to prevent clothing and such from becoming entangled, but they are often missing or nonoperative.

Pudding butt. See *springing*.

Range cubes. See *cattle cubes*.

Ration. A prepared diet for cattle designed to increase weight gain or milk production. A ration can replace all or part of the nutrients obtained from grazing grass. Typical components include hay, grains, legumes, and minerals.

Reactivity. How an animal reacts to humans. Is the animal passive or active?

Regret theory. A theory for making decisions that aims to minimize future regret.

ROPS. Roll-Over Protection System. A hinged bar that can be raised over the driver's seat of a tractor. Prevents the driver from being crushed if the tractor rolls over. This assumes the driver is wearing a seat belt and isn't ejected. In practice, nobody wears the seat belt because they're constantly getting in and out of the tractor.

Rumen. The large chamber on the left flank for initial digestion of grass and hay. The rumen is filled with fluid and bacteria that begin breaking down fibrous material.

Seed-and-feed. A store catering to the needs of local farmers. It sells seed and feed, along with other agricultural supplies.

Service animal. An animal that is individually trained to do work or perform tasks for the benefit of an individual with a disability. See also *emotional support animal.*

Springing. Enlargement of a cow's vulva just prior to calving. When the cow walks, the swollen tissue will bounce back and forth, like a spring. Also called *pudding butt.*

Standing heat. The period of twenty-four hours during the cow's estrous cycle when she can become pregnant. This is the only time she will stand and let the bull mount her.

Tedder. Implement for fluffing fields of cut hay.

Temperament. Inherited tendencies appearing in infancy and continuing throughout life that form the basis of personality.

Three Fs. The three uses of animals: food, fiber, and friends.

Three-point hitch. Mechanical system for attaching implements to a tractor, consisting of three arms sticking out from the rear of the tractor: two below and one up top in a triangular configuration. At the end of each arm, a large bearing accepts the axle from the implement. Using the tractor's hydraulic system, the two lower arms can be raised to lift an implement weighing more than a ton.

Tillage. The breakup of soil for planting. Primary tillage is accomplished with a plow and turns over a foot or more of soil. Secondary tillage uses a harrow or cultivator to even out the top few inches of soil.

Udder drop. When a cow, or heifer, is close to calving, the udder swells and drops below the abdomen. Also called *bagging up.*

Windrow. Hay raked into a row for drying and baling.

Zebu. Generic term for humped cattle derived from the Asian species *Bos indicus*. Also called *Brahma*. Physical characteristics include a fatty hump between the shoulders, a large dewlap, and droopy ears. They are adapted to hot climates. Approximately seventy-five breeds are recognized. They are considered sacred in India.

Acknowledgments

I am lucky to have landed in the orbit of Pastor Ken Peek. There is much that is unwritten about the care of cows, and Ken's wisdom got me out of many jams of my own doing. He and his wife, Boni, have been the most welcoming of neighbors one could imagine, and I am grateful to also call them friends.

Many others in our neck of the woods have also been great sources of support and friendship. Mark and Pam Wallace, our other immediate neighbors, welcomed us too. I will be forever grateful to Mark for planting my pastures that first season and showing me the benefits of a no-till drill. Stacy Patton, who has been raising cattle for years and has far more equipment than I do, has always been there when I needed help, whether moving cows or fixing implements, and so, too, have Dan and Julie Nestor. And thanks to Jerome Newton for his friendship and willingness to sit with the cows.

A special thanks to neighbors Shawn and Sierra Hawkins, who moved to the country about the same time we did. As fellow newbies to farm life, we bonded over our shared experiences in both the joys and challenges of learning to grow food and raise animals.

Across the county, shout-outs to: Melissa Fulton, DVM, and Mary Kate Jobe for doctoring the cows; Chanda Thompson, DVM, for the early vaccinations and castrations; and Tom Crowley, for helping complete strangers who had no business owning cattle get Xena latched on that first day, kicking off this wonderful adventure. Although I got off to a rocky start, the staff at Smith's Seed and Feed are truly wonderful folks, and I now enjoy picking their brains for advice. Thanks to Nicki Halstead, Kim Ash, and Michael Carruth.

Across the rural counties of middle Georgia, Chad Boyce came to the rescue when hooves needed trimming. And in the far southwest corner of the state, Will Harris was gracious in showing me the inner workings of White

Oak Pastures. Thanks to Yaniv Assaf for sharing his MRI data of the cow brain. Such generosity is often rare these days in academia.

Thanks to Michelle Tessler for believing in my passion project and introducing me to Matt Baugher at Harper Horizon, who had me at "Don't let anyone change that title." Meaghan Porter helped me tighten up the prose and moved everything through production in record time.

None of this would have been possible without the support of my family. My children, Helen and Madeline, indulged my schemes from afar, but I owe special gratitude to Kathleen for taking a blind leap of faith in joining me on this journey. My father, Michael Berns, had intended to join us here, but passed without ever seeing the farm. I think about him every time I walk the pastures. I miss you, Dad.

Notes

Chapter 2: Buying Tara

1. *Gone with the Wind*, directed by Victor Fleming (Selznick International Pictures, 1939).
2. Mapsbynik, "Nobody Lives Here: The Nearly 5 Million Census Blocks with Zero Population," https://tumblr.mapsbynik.com/post/82791188950 /nobody-lives-here-the-nearly-5-million-census.
3. Margaret Mitchell, *Gone with the Wind* (New York: Scribner, 2011), 45.

Chapter 3: The Cows Come Home

1. International Miniature Zebu Association, http://imza.name.
2. *Talladega Nights: The Ballad of Ricky Bobby*, directed by Adam McKay (Sony Pictures, 2006).
3. Temple Grandin, *Temple Grandin's Guide to Working with Farm Animals: Safe, Humane Livestock Handling Practices for the Small Farm* (North Adams, MA: Storey Publishing, 2017).

Chapter 4: Cow Brains

1. Gregory Berns, *What It's Like to Be a Dog: And Other Adventures in Animal Neuroscience* (New York: Basic Books, 2017).
2. ASM Mammal Diversity Database, https://www.mammaldiversity.org.
3. Peter F. Cook et al., "Awake Canine FMRI Predicts Dogs' Preference for Praise and Food," *Social Cognitive Affective Neuroscience* 11 (2016): 1853–62, https:// doi.org/10.1101/062703.
4. Ashley Prichard et al., "Awake FMRI Reveals Brain Regions for Novel Word Detection in Dogs," *Frontiers in Neuroscience* 12 (2018): 737, https://doi.org/10 .3389/fnins.2018.00737.
5. Yaniv Assaf et al., "Conservation of Brain Connectivity and Wiring Across the Mammalian Class," *Nature Neuroscience* 23, no. 7 (2020): 805–8, https://doi .org/10.1038/s41593-020-0641-7; Benjamin C. Tendler et al., "The Digital

Brain Bank, an Open Access Platform for Post-Mortem Imaging Datasets,"
Elife 11 (2022): e73153, https://doi.org/10.7554/eLife.73153.

6. Harry J. Jerison, *Evolution of the Brain and Intelligence* (New York: Academic, 1973).

7. Gerard Roth and Ursula Dicke, "Evolution of the Brain and Intelligence," *Trends in Cognitive Science* 9, no. 5 (2014): 250–57, https://doi.org/10.1016/j .tics.2005.03.005.

8. Jean Piaget, *The Construction of Reality in the Child* (Oxon: Routledge, 1954).

9. Evan L. MacLean et al., "The Evolution of Self-Control," *Proceedings of the National Academy of Sciences* 111, no. 20 (2014): E2140–E48, https://doi.org /10.1073/pnas.132353311.

10. Pasko Rakic, "A Small Step for the Cell, a Giant Leap for Mankind: A Hypothesis of Neocortical Expansion During Evolution," *Trends in Neurosciences* 18, no. 9 (1995): 383–88, https://doi.org/10.1016/0166 -2236(95)93934-P.

11. Paul R. Manger et al., "Amplification of Potential Thermogenetic Mechanisms in Cetacean Brains Compared to Artiodactyl Brains," *Scientific Reports* 11, no. 1 (2021): 1–15, https://doi.org/10.1038/s41598-021-84762-0.

12. Rodrigo S. Kazu et al., "Cellular Scaling Rules for the Brain of Artiodactyla Include a Highly Folded Cortex with Few Neurons," *Frontiers in Neuroanatomy* 8 (2014): 128, https://doi.org/10.3389/fnana.2014.00128.

13. Praneshri Pillay and Paul R. Manger, "Order-Specific Quantitative Patterns of Cortical Gyrification," *European Journal of Neuroscience* 25 (2007): 2705–12, https://doi.org/10.1111/j.1460-9568.2007.05524.x; Karl Zilles, Nicola Palomero-Gallagher, and Katrin Amunts, "Development of Cortical Folding During Evolution and Ontogeny," *Trends in Neurosciences* 36, no. 5 (2013): 275–84, https://doi.org/10.1016/j.tins.2013.01.006.

14. Ehsan Tadayon, Alvaro Pascual-Leone, and Emiliano Santarnecchi, "Differential Contribution of Cortical Thickness, Surface Area, and Gyrification to Fluid and Crystallized Intelligence," *Cerebral Cortex* 30, no. 1 (2019): 215–25, https://doi.org/10.1093/cercor/bhz082.

Chapter 5: Cowpuppy

1. Marie-Antonine Finkemeier, Jan Langbein, and Birger Puppe, "Personality Research in Mammalian Farm Animals: Concepts, Measures, and Relationships to Welfare," *Frontiers in Veterinary Science* 5 (2018): 131, https:// doi.org/10.3389/fvets.2018.00131; Marie I. Kaiser and Caroline Müller,

"What Is an Animal Personality?" *Biology & Philosophy* 36, no. 1 (2021): 1–25, https://doi.org/10.1007/s10539-020-09776-w.

2. Oliver P. John, "The 'Big Five' Factor Taxonomy: Dimensions of Personality in the Natural Language and in Questionnaires," in *Handbook of Personality: Theory and Research*, edited by L. A. Pervin (New York: Guilford Press, 1990), 66–100.

3. Samuel D. Gosling and Oliver P. John, "Personality Dimensions in Nonhuman Animals: A Cross-Species Review," *Current Directions in Psychological Science* 8, no. 3 (1999): 69–75, https://doi.org/10.1111/1467-8721.00017.

4. Jenny Gibbons, Alistair Lawrence, and Marie Haskell, "Responsiveness of Dairy Cows to Human Approach and Novel Stimuli," *Applied Animal Behaviour Science* 116, no. 2–4 (2009): 163–73, https://doi.org/10.1016/j.applanim.2008 .08.009; Aline C. Sant'Anna and Mateus J. R. Paranhos da Costa, "Validity and Feasibility of Qualitative Behavior Assessment for the Evaluation of Nellore Cattle Temperament," *Livestock Science* 157, no. 1 (2013): 254–62, https://doi.org/10.1016/j.livsci.2013.08.004.

Chapter 6: Grass

1. Muhammad Haroon Sarwar et al., "The Importance of Cereals (Poaceae: Gramineae) Nutrition in Human Health: A Review," *Journal of Cereals and Oilseeds* 4 (2013): 32–35, https://doi.org/10.5897/JCO12.023.

2. "Water Requirements for Beef Cattle," University of Nebraska–Lincoln Institute of Agriculture and Natural Resources, July 2015, https://beef.unl .edu/water-requirements-for-beef-cattle.

Chapter 7: Cow Love

1. Turid Rugaas, *On Talking Terms with Dogs: Calming Signals* (Wenatchee, WA: Dogwise Publishing, 1997).

2. S. Sato, S. Sako, and A. Maeda, "Social Licking Patterns in Cattle (*Bos Taurus*): Influence of Environmental and Social Factors," *Applied Animal Behaviour Science* 32 (1991): 3–12, https://doi.org/10.1016/S0168-1591(05)80158-3.

3. Shusuke Sato, Keijiro Tarumizu, and Koichi Hatae, "The Influence of Social Factors on Allogrooming in Cows," *Applied Animal Behaviour Science* 38 (1993): 235–44, https://doi.org/10.1016/0168-1591(93)90022-H.

4. Viktor Reinhardt and Annie Reinhardt, "Cohesive Relationships in a Cattle Herd (*Bos Indicus*)," *Behaviour* 77, no. 3 (1981): 121–51, https://www.jstor.org /stable/4534115.

5. Larry Young and Brian Alexander, *The Chemistry Between Us: Love, Sex, and the Science of Attraction* (New York: Current, 2012); C. Sue Carter, "Oxytocin and Love: Myths, Metaphors and Mysteries," *Comprehensive Psychoneuroendocrinology* (2021): 100107, https://doi.org/10.1016/j.cpnec.2021.100107.
6. Stephanie Lurzel et al., "Salivary Oxytocin in Pigs, Cattle, and Goats During Positive Human-Animal Interactions," *Psychoneuroendocrinology* 115 (2020): 104636, https://doi.org/10.1016/j.psyneuen.2020.104636.
7. Rosamund Young, *The Secret Life of Cows* (New York: Penguin Books, 2018).
8. Claudia Schmied et al., "Stroking of Different Body Regions by a Human: Effects on Behaviour and Heart Rate of Dairy Cows," *Applied Animal Behaviour Science* 109 (2008): 25–38, https://doi.org/10.1016/j.applanim.2007.01.013.
9. Temple Grandin, *Temple Grandin's Guide to Working with Farm Animals: Safe, Humane Livestock Handling Practices for the Small Farm* (North Adams, MA: Storey Publishing, 2017).

Chapter 8: Dogs Versus Cows

1. Erik Axelsson et al., "The Genomic Signature of Dog Domestication Reveals Adaptation to a Starch-Rich Diet," *Nature* 495, no. 7441 (2013): 360–64, https://doi.org/10.1038/nature11837.
2. Peter F. Cook et al., "Awake Canine FMRI Predicts Dogs' Preference for Praise and Food," *Social Cognitive Affective Neuroscience* 11 (2016): 1853–62, https://doi.org/10.1101/062703.
3. Clive D. L. Wynne, *Dog Is Love: Why and How Your Dog Loves You* (New York: Houghton Mifflin, 2019).

Chapter 9: Through a Cow's Eyes

1. Martin S. Banks et al., "Why Do Animal Eyes Have Pupils of Different Shapes?" *Science Advances* 1 (2015): e1500391, https://doi.org/10.1126/sciadv .1500391.
2. "Color Blindness," National Eye Institute, accessed February 22, 2024, https:// www.nei.nih.gov/learn-about-eye-health/eye-conditions-and-diseases/color -blindness.
3. Leo M. Hurvich and Dorothea Jameson, "An Opponent-Process Theory of Color Vision," *Psychological Review* 64, no. 6p1 (1957): 384, https://doi.org/10 .1037/h0041403.
4. Gerald H. Jacobs, Jess F. Deegan II, and Jay Neitz, "Photopigment Basis for Dichromatic Color Vision in Cows, Goats, and Sheep," *Visual Neuroscience* 15, no. 3 (1998): 581–84, https://doi.org/10.1017/S0952523898153154; C. J. C.

Phillips and C. A. Lomas, "The Perception of Color by Cattle and Its Influence on Behavior," *Journal of Dairy Science* 84, no. 4 (2001): 807–13, https://doi.org /10.3168/jds.S0022-0302(01)74537-7.

5. Temple Grandin, *Temple Grandin's Guide to Working with Farm Animals: Safe, Humane Livestock Handling Practices for the Small Farm* (North Adams, MA: Storey Publishing, 2017).

6. Phillips and Lomas, "The Perception of Color by Cattle and Its Influence on Behavior."

7. J. A. Riol et al., "Colour Perception in Fighting Cattle," *Applied Animal Behaviour Science* 23 (1989): 199–206, https://doi.org/10.1016/0168 -1591(89)90110-X.

8. Hans Brettel, Francoise Vienot, and John D. Mollon, "Computerized Simulation of Color Appearance for Dichromats," *Journal of the Optical Society of America A* 14, no. 10 (1997): 2647–55, https://doi.org/10.1364/JOSAA .14.002647. For simulations, see DaltonLens, "Online Color Blindness Simulators," accessed January 16, 2024, https://daltonlens.org/colorblindness -simulator.

Chapter 10: Personal Space

1. Robert Sommer, "Studies in Personal Space," *Sociometry* 22, no. 3 (1959): 247 –60, https://doi.org/10.2307/2785668.

2. Sarah N.Sells et al., "Evidence of Economical Territory Selection in a Cooperative Carnivore," *Proceedings of the Royal Society B: Biological Sciences* 288, no. 1946 (2021): 20210108, https://doi.org/10.1098/rspb.2021.0108.

3. Samuel B. Hunley and Stella F. Lourenco, "What Is Peripersonal Space? An Examination of Unresolved Empirical Issues and Emerging Findings," *Wiley Interdisciplinary Reviews: Cognitive Science* 9, no. 6 (2018): e1472, https://doi .org/10.1002/wcs.1472.

4. Justine Cléry et al., "Neuronal Bases of Peripersonal and Extrapersonal Spaces, Their Plasticity and Their Dynamics: Knowns and Unknowns," *Neuropsychologia* 70 (2015): 313–26, https://doi.org/10.1016/j .neuropsychologia.2014.10.022.

5. F. de Vignemont and G. D. Iannetti, "How Many Peripersonal Spaces?" *Neuropsychologia* 70 (2015): 327–34, https://doi.org/10.1016/j .neuropsychologia.2014.11.018.

6. Michael S. A. Graziano and Dylan F. Cooke, "Parieto-Frontal Interactions, Personal Space, and Defensive Behavior," *Neuropsychologia* 44 (2006): 845–59, https://doi.org/10.1016/j.neuropsychologia.2005.09.011.

7. Daphne J. Holt et al., "Neural Correlates of Personal Space Intrusion," *Journal of Neuroscience* 34, no. 12 (2014): 4123–34, https://doi.org/10.1523/JNEUROSCI.0686-13.2014.

8. Temple Grandin, *Temple Grandin's Guide to Working with Farm Animals: Safe, Humane Livestock Handling Practices for the Small Farm* (North Adams, MA: Storey Publishing, 2017).

9. Mette S. Herskin, Lene Munksgaard, and Jan Ladewig, "Effects of Acute Stressors on Nociception, Adrenocortical Responses and Behavior of Dairy Cows," *Physiology & Behavior* 83, no. 3 (2004): 411–20, https://doi.org/10.1016/j.physbeh.2004.08.027.

Chapter 12: Stockmanship

1. Temple Grandin, *Temple Grandin's Guide to Working with Farm Animals: Safe, Humane Livestock Handling Practices for the Small Farm* (North Adams, MA: Storey Publishing, 2017).

2. Whit Hibbard and Matt Barnes, "Stockmanship and Range Management," *Stockmanship Journal* 5, no. 2 (2016): 1–15.

3. "Bud Williams Stockmanship and Livestock Marketing," accessed April 5, 2023, https://stockmanship.com.

4. Temple Grandin, "The Visual, Auditory, and Physical Environment of Livestock Handling Facilities and Its Effect on Ease of Movement of Cattle, Pigs, and Sheep," *Frontiers in Animal Science* 2 (2021): 744207, https://doi.org/10.3390/ani8120236.

5. Steve Cote, *Manual of Stockmanship* (Salt Lake City, UT: Hudson Printing Company, 2019).

6. Mette S. Herskin, Lene Munksgaard, and Jan Ladewig, "Effects of Acute Stressors on Nociception, Adrenocortical Responses and Behavior of Dairy Cows," *Physiology & Behavior* 83, no. 3 (December 2004): 411–20, https://doi.org/10.1016/j.physbeh.2004.08.027; Carlos E. Hernandez et al., "Time Lag Between Peak Concentrations of Plasma and Salivary Cortisol Following a Stressful Procedure in Dairy Cattle," *Acta Veterinaria Scandinavica* 56 (2014): 1–8, https://doi.org/10.1186/s13028-014-0061-3.

Chapter 13: Cow Cam

1. Emma Ternman et al., "First-Night Effect on Sleep Time in Dairy Cows," *PLoS ONE* 13, no. 4 (2018): e0195593, https://doi.org/10.1371/journal.pone.0195593.

2. Steve Cote, *Manual of Stockmanship.*

3. Iain D. Couzin, "Collective Cognition in Animal Groups," *Trends in Cognitive Sciences* 13, no. 1 (2009): 36–43, https://doi.org/10.1016/j.tics.2008.10.002.

4. Jolle W. Jolles, Andrew J. King, and Shaun S. Killen, "The Role of Individual Heterogeneity in Collective Animal Behaviour," *Trends in Ecology & Evolution* 35, no. 3 (2020): 278–91, https://doi.org/10.1016/j.tree.2019.11.001.

5. I used a convolutional neural net called Detectron2, a program developed by Meta and released for free public use.

Chapter 14: Ricky Bobby Holds a Grudge

1. K. Rosenberger et al., "Performance of Goats in a Detour and a Problem -Solving Test Following Long-Term Cognitive Test Exposure," *Royal Society Open Science* 8, no. 10 (2021): 210656, https://doi.org/10.1098/rsos.210656.

2. José Z. Abramson et al., "Spatial Perseveration Error by Alpacas (Vicugna Pacos) in an A-Not-B Detour Task," *Animal Cognition* 21 (2018): 433–39, https://doi.org/10.1007/s10071-018-1170-6.

3. Alvaro Lopez Caicoya et al., "Comparative Cognition in Three Understudied Ungulate Species: European Bison, Forest Buffalos and Giraffes," *Frontiers in Zoology* 18, no. 1 (2021): 30, https://doi.org/10.1186/s12983-021-00417-w.

4. Masahiko Hirata, Chihiro Tomita, and Karin Yamada, "Use of a Maze Test to Assess Spatial Learning and Memory in Cattle: Can Cattle Traverse a Complex Maze?" *Applied Animal Behaviour Science* 180 (2016): 18–25, https://doi.org/10.1016/j.applanim.2016.04.004.

5. Can Kabadayi, Katarzyna Bobrowicz, and Mathias Osvath, "The Detour Paradigm in Animal Cognition," *Animal Cognition* 21 (2018): 21–35, https://doi.org/10.1007/s10071-017-1152-0.

6. Marjorie Coulon et al., "Cattle Discriminate Between Familiar and Unfamiliar Conspecifics by Using Only Head Visual Cues," *Animal Cognition* 14 (2011): 279–90, https://doi.org/10.1007/s10071-010-0361-6.

7. Endel Tulving, "Episodic Memory, from Mind to Brain," *Annual Review of Psychology* 53 (2002): 1–25, https://doi.org/10.1146/annurev.psych.53 .100901.135114.

8. E. Dere et al., "The Case for Episodic Memory in Animals," *Neuroscience and Biobehavioral Reviews* 30 (2006): 1206–24, https://doi.org/10.1016/j.neubiorev .2006.09.005.

9. Danielle Panoz-Brown et al., "Replay of Episodic Memories in the Rat," *Current Biology* 28, no. 10 (2018): 1628–34.e7, https://doi.org/10.1016/j.cub .2018.04.006.

10. Jonathan D. Crystal and Thomas Suddendorf, "Episodic Memory in

Nonhuman Animals?" *Current Biology* 29 (2019): R1291–R95, https://doi.org /10.1016/j.cub.2019.10.045.

Chapter 16: Playtime

1. Gordon M. Burghardt, *The Genesis of Animal Play* (Cambridge, MA: MIT Press, 2005).
2. Marc Bekoff, "The Development of Social Interaction, Play, and Metacommunication in Mammals: An Ethological Perspective," *Quarterly Review of Biology* 47, no. 4 (1972): 412–34, https://www.jstor.org/stable /2820738.
3. Lisa A. Parr et al., "Classifying Chimpanzee Facial Expressions Using Muscle Action," *Emotion* 7 (2007): 172–81, https://doi.org/10.1037/1528-3542.7.1 .172.
4. V. Reinhardt, F. M. Mutiso, and A. Reinhardt, "Social Behaviour and Social Relationships Between Female and Male Prepubertal Bovine Calves (*Bos Indicus*)," *Applied Animal Ethology* 4 (1978): 43–54, https://doi.org/10.1016 /0304-3762(78)90092-5.
5. Herbert Spencer, *Principles of Psychology* (New York: Appleton, 1872).
6. William Lauder Lindsay, *Mind in the Lower Animals: In Health and Disease* (London: C. Kegan Paul & Co., 1879).
7. Burghardt, *The Genesis of Animal Play*, 30.
8. Harvey A. Carr, "The Survival Values of Play," in *Investigations of the Department of Psychology and Education of the University of Colorado*, edited by Arthur Allin (Boulder, CO: University of Colorado, 1902), 1–47.
9. A. Brownlee, "Play in Domestic Cattle in Britain: An Analysis of Its Nature," *British Veterinary Journal* 110 (1954): 48–68, https://doi.org/10.1016/S0007 -1935(17)50529-1.
10. John A. Byers and Curt Walker, "Refining the Motor Training Hypothesis for the Evolution of Play," *American Naturalist* 146, no. 1 (1995): 25–40, https:// www.jstor.org/stable/2463035.
11. Burghardt, *The Genesis of Animal Play,* 161.
12. Martin B. Main, Floyd W. Weckerly, and Vernon C. Bleich, "Sexual Segregation in Ungulates: New Directions for Research," *Journal of Mammalogy* 77, no. 2 (1996): 449–61, https://doi.org/10.2307/1382821.

Chapter 17: Mirror, Mirror

1. Gordon G. Gallup and James R. Anderson, "Self-Recognition in Animals: Where Do We Stand 50 Years Later? Lessons from Cleaner Wrasse and Other

Species," *Psychology of Consciousness: Theory, Research, and Practice* 7, no. 1 (2020): 46–58, https://doi.org/10.1037/cns0000206.

2. Gordon G. Gallup, "Mirror-Image Stimulation," *Psychological Bulletin* 70, no. 6 (1968): 782–93, https://doi.org/10.1037/h0026777.

3. Monkeys can pass the mirror test if given much more intensive training that includes tactile stimulation of the mark. See Liangtang Chang et al., "Mirror-Induced Self-Directed Behaviors in Rhesus Monkeys after Visual -Somatosensory Training," *Current Biology* 25, no. 2 (2015): 212–17, https://doi.org/10.1016/j.cub.2014.11.016.

4. Alexandra Horowitz, "Smelling Themselves: Dogs Investigate Their Own Body Odours Longer When Modified in an 'Olfactory Mirror' Test," *Behavioural Processes* 143 (2017): 17–24, https://doi.org/10.1016/j.beproc.2017.08.001.

5. Frans B. M. de Waal, "Fish, Mirrors, and a Gradualist Perspective on Self -Awareness," *PLoS Biology* 17, no. 2 (2019): e3000112, https://doi.org/10.1371/journal.pbio.3000112.

6. S. D. McBride, N. Perentos, and A. J. Morton, "Understanding the Concept of a Reflective Surface: Can Sheep Improve Navigational Ability Through the Use of a Mirror?" *Animal Cognition* 18 (2015): 361–71 (2015), https://doi.org/10.1007/s10071-014-0807-3.

7. McBride et al., "Understanding the Concept of a Reflective Surface."

8. Paolo Baragli et al., "Are Horses Capable of Mirror Self-Recognition? A Pilot Study," *PLoS ONE* 12, no. 5 (2017): e0176717, https://doi.org/10.1371/journal.pone.0176717.

9. Donald M. Broom, Hilana Sena, and Kiera L. Moynihan, "Pigs Learn What a Mirror Image Represents and Use It to Obtain Information," *Animal Behaviour* 78, no. 5 (2009): 1037–41, https://doi.org/10.1016/j.anbehav.2009.07.027.

10. Sonja Hillemacher et al., "Roosters Do Not Warn the Bird in the Mirror: The Cognitive Ecology of Mirror Self-Recognition," *PLoS ONE* 18, no. 10 (2023): e0291416, https://doi.org/10.1371/journal.pone.0291416.

11. Carol A. K. Piller et al., "Effects of Mirror-Image Exposure on Heart Rate and Movement of Isolated Heifers," *Applied Animal Behaviour Science* 63, no. 2 (1999): 93–102, https://doi.org/10.1016/S0168-1591(99)00010-6.

Chapter 18: What Does the Cow Say?

1. "Cow Sounds," Omniglot, accessed July 3, 2023, https://omniglot.com/language/animalsounds/cows.htm.

2. Temple Grandin, "Cattle Vocalizations Are Associated with Handling and

Equipment Problems at Beef Slaughter Plants," *Applied Animal Behaviour Science* 71, no. 3 (2001): 191–201, https://doi.org/10.1016/S0168 -1591(00)00179-9.

3. M. A. Schnaider et al., "Vocalization and Other Behaviors as Indicators of Emotional Valence: The Case of Cow-Calf Separation and Reunion in Beef Cattle," *Journal of Veterinary Behavior* 49 (2022): 28–35, https://doi.org/10 .1016/j.jveb.2021.11.011.

4. Daniel M. Weary and Beverly Chua, "Effects of Early Separation on the Dairy Cow and Calf: 1. Separation at 6 H, 1 Day and 4 Days after Birth," *Applied Animal Behaviour Science* 69, no. 3 (2000): 177–88, https://doi.org/10.1016/j .jveb.2021.11.011.

5. Jon M. Watts and Joseph M. Stookey, "Vocal Behaviour in Cattle: The Animal's Commentary on Its Biological Processes and Welfare," *Applied Animal Behaviour Science* 67 (2000): 15–33, https://doi.org/10.1016/S0168 -1591(99)00108-2; Gerhard Manteuffel, Birger Puppe, and Peter C. Schon, "Vocalization of Farm Animals as a Measure of Welfare," *Applied Animal Behaviour Science* 88 (2004): 163–82, https://doi.org/10.1016/j.applanim .2004.02.012.

6. Christine H. Barfield, Zuleyma Tang-Martinez, and Jill M. Trainer, "Domestic Calves (*Bos Taurus*) Recognize Their Own Mothers by Auditory Cues," *Ethology* 97, no. 4 (1994): 257–64, https://doi.org/10.1111/j.1439-0310 .1994.tb01045.x.

7. Singh Yajuvendra et al., "Effective and Accurate Discrimination of Individual Dairy Cattle Through Acoustic Sensing," *Applied Animal Behaviour Science* 146, no. 1–4 (2013): 11–18, https://doi.org/10.1016/j.applanim.2013.03.008.

8. Raymond D. Kent and Houri K. Vorperian, "Static Measurements of Vowel Formant Frequencies and Bandwidths: A Review," *Journal of Communication Disorders* 74 (2018): 74–97, https://doi.org/10.1016/j.jcomdis.2018.05.004.

9. Monica Padilla de la Torre et al., "Acoustic Analysis of Cattle (*Bos Taurus*) Mother-Offspring Contact Calls from a Source-Filter Theory Perspective," *Applied Animal Behaviour Science* 163 (2015): 58–68, https://doi.org/10.1016/j .applanim.2014.11.017.

10. Alexandra Green et al., "Vocal Individuality of Holstein-Friesian Cattle Is Maintained Across Putatively Positive and Negative Farming Contexts," *Scientific Reports* 9, art. 18468 (2019), https://doi.org/10.1038/s41598-019 -54968-4.

11. Suresh Neethirajan, Inonge Reimert, and Bas Kemp, "Measuring Farm

Animal Emotions—Sensor-Based Approaches," *Sensors* 21, no 2. (2021): 553, https://doi.org/10.3390/s21020553.

12. Dagmar M. Schuller and Bjorn W. Schuller, "A Review on Five Recent and Near-Future Developments in Computational Processing of Emotion in the Human Voice," *Emotion Review* 13, no. 1 (2021): 44–50, https://doi.org/10 .1177/1754073919898.

13. Marina Scheumann et al., "The Voice of Emotion Across Species: How Do Human Listeners Recognize Animals' Affective States?" *PLoS ONE* 9, no. 3 (2014): e91192, https://doi.org/10.1371/journal.pone.0091192.

14. Peter Pongracz, Csaba Molnar, and Adam Miklosi, "Acoustic Parameters of Dog Barks Carry Emotional Information for Humans," *Applied Animal Behaviour Science* 100 (2006): 228–40, https://doi.org/10.1016/j.applanim .2005.12.004.

15. Eugene S. Morton, "On the Occurrence and Significance of Motivation-- Structural Rules in Some Bird and Mammal Sounds," *The American Naturalist* 111, no. 981 (1977): 855–69, https://www.jstor.org/stable/2460385.

16. Csaba Molnar et al., "Classification of Dog Barks: A Machine Learning Approach," *Animal Cognition* 11 (2008): 389–400, https://doi.org/10.1007 /s10071-007-0129-9.

17. Alexandra C. Green et al., "Vocal Production in Postpartum Dairy Cows: Temporal Organization and Association with Maternal and Stress Behaviors," *Journal of Dairy Science* 104 (2021): 826–38, https://doi.org/10.3168/jds.2020 -18891.

18. Larry W. Richardson et al., "Acoustics of White-Tailed Deer (*Odocoileus Virginianus*)," *Journal of Mammalogy* 64, no. 2 (1983): 245–52, https://doi.org /10.2307/1380554.

19. Sune Borkfelt, "What's in a Name?—Consequences of Naming Non-Human Animals," *Animals* 1 (2011): 116–25, https://doi.org/10.3390/ani1010116.

20. Juliane Kaminski, Linda Schulz, and Michael Tomasello, "How Dogs Know When Communication Is Intended for Them," *Developmental Science* 15, no. 2 (2012): 222–32, https://doi.org/10.1111/j.1467-7687.2011.01120.x.

21. K. Uetake, J. F. Hurnik, and L. Johnson, "Effect of Music on Voluntary Approach of Dairy Cows to an Automatic Milking System," *Applied Animal Behaviour Science* 53, no. 3 (1997): 175–82, https://doi.org/10.1016/S0168 -1591(96)01159-8; Patrycja Ciborowska, Monika Michalczuk, and Damian Bien, "The Effect of Music on Livestock: Cattle, Poultry and Pigs," *Animals* 11, no. 12 (2021): 3572, https://doi.org/10.3390/ani11123572.

Chapter 19: Cow Culture

1. Andrew Whiten, "The Burgeoning Reach of Animal Culture," *Science* 372, no. 6537 (2021): eabe6514, https://doi.org/10.3758/s13423-022-02236-4.
2. J. Fisher and R. A. Hinde, "The Opening of Milk Bottles by Birds," *British Birds* 42 (1949): 347–57.
3. Peter Marler and Miwako Tamura, "Culturally Transmitted Patterns of Vocal Behavior in Sparrows," *Science* 146 (1964): 1483–86, https://doi.org/10.1126/science.146.3650.1483.
4. Sylvain Alem et al., "Associative Mechanisms Allow for Social Learning and Cultural Transmission of String Pulling in an Insect," *PLoS Biology* 14, no. 10 (2016): e1002564, https://doi.org/10.1371/journal.pbio.1002564.
5. Isabelle Veissier, "Observational Learning in Cattle," *Applied Animal Behaviour Science* 35, no. 3 (1993): 235–43, https://doi.org/10.1016/0168-1591(93)90139-G; Johanna Stenfelt et al., "Dairy Cows Did Not Rely on Social Learning Mechanisms When Solving a Spatial Detour Task," *Frontiers in Veterinary Science* 9 (2022): 956559, https://doi.org/10.3389/fvets.2022.956559.
6. Hamideh Keshavarzi et al., "Virtual Fence Responses Are Socially Facilitated in Beef Cattle," *Frontiers in Veterinary Science* 7 (2020): 543158, https://doi.org/10.3389/fvets.2020.543158.

Chapter 20: Pasture Ornaments

1. "Cattle Inventory," USDA, 2021, https://www.nass.usda.gov/Surveys/Guide_to_NASS_Surveys/Cattle_Inventory.
2. Will Harris, *A Bold Return to Giving a Damn: One Farm, Six Generations, and the Future of Food* (New York: Viking, 2023).
3. "Feedlot Operations: Why It Matters Where Your Grain-Finished Beef Was Produced," NRDC, November 13, 2014, accessed July 26, 2023, https://www.nrdc.org/resources/feedlot-operations-why-it-matters.
4. "Animal Feeding Operations (AFOs)," United States Environmental Protection Agency, accessed July 26, 2023, https://www.epa.gov/npdes/animal-feeding-operations-afos.
5. D. Lee Miller and Gregory Muren, "CAFOs: What We Don't Know Is Hurting Us," NRDC, September 23, 2019, https://www.nrdc.org/resources/cafos-what-we-dont-know-hurting-us.
6. Andy Reisinger and Harry Clark, "How Much Do Direct Livestock Emissions Actually Contribute to Global Warming?" *Global Change Biology* 24, no. 4 (2017): 1749–61, https://doi.org/10.1111/gcb.13975.

7. Mario Herrero et al., "Greenhouse Gas Mitigation Potentials in the Livestock Sector," *Nature Climate Change* 6, no. 5 (2016): 452–61, https://doi.org/10.1038/nclimate2925.

8. Gregory S. Okin, "Environmental Impacts of Food Consumption by Dogs and Cats," *PLoS ONE* 12, no. 8 (2017): e0181301, https://doi.org/10.1371/journal.pone.0181301.

9. "Animal Welfare," World Organization for Animal Health, accessed April 9, 2024, https://www.woah.org/en/what-we-do/animal-health-and-welfare/animal-welfare/.

10. "Top 9 Meat Packing Plants in the U.S.," IndustrySelect, July 20, 2022, accessed July 27, 2023, https://www.industryselect.com/blog/the-largest-meatpacking-plants-in-the-us.

11. Yi-Yuan Tang et al., "Frontal Theta Activity and White Matter Plasticity Following Mindfulness Meditation," *Current Opinion in Psychology* 28 (2019): 294–97, https://doi.org/10.1016/j.copsyc.2019.04.004.

12. "About Us," Georgia Cattlewomen's Association, accessed August 2, 2023, https://www.georgiacattlewomen.org/about-us.

13. Hal Herzog, *Some We Love, Some We Hate, Some We Eat: Why It's So Hard to Think Straight About Animals* (New York: HarperCollins, 2010).

14. Florence Nightingale, *Notes on Nursing: What It Is, and What It Is Not* (London: Harrison and Sons, 1859), 103.

15. Yasmin El-Beih, "Is Cow Hugging the World's New Wellness Trend?" BBC Reel, October 9, 2020, https://www.bbc.com/travel/article/20201008-is-cow-hugging-the-worlds-new-wellness-trend.

16. Aubrey H. Fine, Alan M. Beck, and Zenithson Ng, "The State of Animal-Assisted Interventions: Addressing the Contemporary Issues That Will Shape the Future," *International Journal of Environmental Research and Public Health* 16, no. 20 (2019): 3997, https://doi.org/10.3390/ijerph16203997; Marguerite E. O'Haire, Noemie A. Guerin, and Alison C. Kirkham, "Animal-Assisted Intervention for Trauma: A Systematic Literature Review," *Frontiers in Psychology* 6 (2015): 1121, https://doi.org/10.3389/fpsyg.2015.01121.

17. Nancy Parish-Plass, "Animal-Assisted Therapy with Children Suffering from Insecure Attachment Due to Abuse and Neglect: A Method to Lower the Risk of Intergenerational Transmission of Abuse?" *Clinical Child Psychology and Psychiatry* 13, no. 1 (2008): 7–30, https://doi.org/10.1177/135910450708.

Appendix A: A Brief History of the Cow

1. Catrin Rutland, *The Cow: A Natural & Cultural History* (Princeton, NJ: Princeton University Press, 2021).
2. The caves were closed to the public, but a virtual tour was available. "Lascaux Cave," Ministère de la Culture, Paris, France, https://archeologie.culture.gouv.fr/lascaux/en.
3. T. Van Vuure, "History, Morphology and Ecology of the Aurochs (*Bos Primigenius*)," *Lutra: Journal of the Dutch Mammal Society* 45 (2002): 1–45.
4. R. Schafberg and H. H. Swalve, "The History of Breeding for Polled Cattle," *Livestock Science* 179 (2015): 54–70, https://doi.org/10.1016/j.livsci.2015.05.017.
5. J. E. Aldersey et al., "Understanding the Effects of the Bovine Polled Variants," *Animal Genetics* 51 (2020): 166–76, https://doi.org/10.1111/age.12915.
6. "Our Mission," The Livestock Conservancy, https://livestockconservancy.org.
7. The most famous sculpture, the Winged Bull of Nineveh, was destroyed in 2015.
8. "List of Works by Vincent van Gogh," Wikipedia, accessed October 8, 2023, https://en.wikipedia.org/wiki/List_of_works_by_Vincent_van_Gogh#.
9. "The Cattle Industry in the American West," History on the Net, accessed October 9, 2023, https://www.historyonthenet.com/american-west-the-cattle-industry.
10. Timothy Beck Werth, "Meet Cool Whip, the 1,700-Pound 'Sweetie Pie' Bucking for Rodeo Glory," *GQ Sports*, August 29, 2023, https://www.gq.com/story/cool-whip-rodeo-bull-profile.
11. "Nation's Urban and Rural Populations Shift Following 2020 Census," United States Census Bureau, accessed October 9, 2023, https://www.census.gov/newsroom/press-releases/2022/urban-rural-populations.html.

About the Author

Gregory Berns is a professor of psychology at Emory University, where he directs the Center for Neuropolicy and the Facility for Education and Research in Neuroscience. He is the author of several books, including *What It's Like to Be a Dog* and the *New York Times* and *Wall Street Journal* bestseller *How Dogs Love Us*. He and his wife live on a farm near Atlanta, Georgia, with several dogs, chickens, and some very special cows.